NORTHEASTERN TIGER BEETLES

A Field Guide to Tiger Beetles of New England and Eastern Canada

Jonathan G. Leonard
Ross T. Bell

CRC Press
Boca Raton   London   New York   Washington, D.C.

Library of Congress Cataloging-in-Publication Data

Leonard, Jonathan G.
   Northeastern tiger beetles: a field guide to tiger beetles of New England and eastern Canada/
   by Jonathan Leonard and Ross T. Bell.
      p.   cm.
   Includes bibliographical references and index.
   ISBN 0-8493-1915-3
      1. Tiger beetles—New England—Identification. 2. Tiger beetles—Canada,
   Easern—Identification. I. Bell, Ross T. II. Title.
   QL598.C56L46 1998
   595.76'2—dc21
   for Library of Congress

                                               98-41599
                                                  CIP

International Standard Book Number 0-8493-1915-3
Library of Congress Card Number 98-41599
Printed in the United States of America 1 2 3 4 5 6 7 8 9 0
Printed on acid-free paper

PREFACE

Jonathan Leonard noticed his first tiger beetle over 30 years ago on stretches of beach along the Atlantic coast. During long hikes on the barrier beaches of Long Island and Cape Cod, Jonathan noticed tiger beetles flying up from the sand in front of his feet and landing 10 to 15 feet away.

Ross Bell's introduction to tiger beetles was over 50 years ago. As a beginning beetle collector he was attracted by the brilliant green *Cicindela sexguttata* along woodland paths in eastern Illinois. Later, as a college student, Ross went on field trips to the Indiana dunes region at the south end of Lake Michigan, where he saw the segregation of several species of tiger beetles in particular habitats, as first described by Victor Shelford. Ross had the great privilege of knowing Dr. Shelford personally and visited the area with him.

We both discovered that tiger beetles are beautiful when viewed under a magnifying glass or hand lens. Many have striking metallic colors and white patterns or "maculations" on their elytra or wing covers. These maculations are variations of a common theme, yet vary from species to species. On a microscopic scale, the colors of tiger beetles are caused by thin melanin layers in the cuticule that refract and reflect light of different wavelengths to different degrees, causing some wavelengths to be canceled out (Hadley 1986; Knisley and Schultz 1997). The shade of these structural colors often changes when viewed from different angles. In addition, there are tiny cup-shaped hexagonal pits or alveoli that add texture to the tiger beetle elytron color (Fig. 1).

We learned that many species of tiger beetle have a preferred habitat; some species are generalists and some are only found in specific places. Most species are found in sandy, temporary habitats caused by natural disturbances such as spring floods and storms. For example, the rare and beautiful puritan tiger beetle, *Cicindela puritana*, is found in the Northeast only along the sandy shores and eroding banks of the Connecticut River. Years ago, spring floods from the melting of deep winter snows caused the river to erode down through ancient sand deposits causing the present sand bluffs. Flood control dams built along the tributaries of the Connecticut River after World War II have controlled the spring floods. Riverbanks were stabilized as towns like Springfield, MA and Hartford, CT grew. Since then, the eroding sand banks and sandbars have begun to disappear by being overgrown with vegetation. As the temporary beach habitat has become less common, so has *C. puritana*, now federally listed as an endangered species (U.S. Fish and Wildlife Service 1993a). Therefore, periodic disturbance is an important factor in maintaining populations of some tiger beetle species.

Some tiger beetles in New England are in danger of becoming extinct. For example, *C. puritana* appears to be extinct in Vermont and New Hampshire, but is still present in extremely reduced populations in Massachusetts and Connecticut. *Cicindela dorsalis dorsalis*, the northeastern beach tiger beetle, was once very common on our Atlantic shore beaches in southern New England. Near the turn of the last century there were reports of finding them in "great swarms" in July (Leng 1902: 161). Now, with the increase in foot, jeep, and beach-buggy traffic, the fragile larvae have little chance of avoiding a crushing death or at least a drastic modification of their microhabitat (Schultz 1988; Knisley and Schultz 1997). Only two small populations of this beautiful beach tiger beetle are now known in our area along stretches of beach in Massachusetts.

We have written this field guide to make the identification of tiger beetles easier for the amateur naturalist. We hope the study of tiger beetle ecology becomes a popular hobby, much as bird watching became popular after field guides were published. It is our hope that, once people become more aware of tiger beetles and their habitats, these creatures and the fragile areas they live in will become better managed and preserved.

Much of the information in this book is taken from specialty journals and books that are not easy to find unless you have access to a university research library. We have listed references at the end of this book for those who wish more information or would like to read the original sources. References in the text are given in parentheses as (author /date).

As you become more interested in learning about tiger beetles you may want to subscribe to the quarterly journal *Cicindela*. Subscription to this marvelous journal is an exceptional bargain at $7.00 per year. Subscription requests should be addressed to the senior editor: Ronald L. Huber, 4637 W. 69th Terrace, Prairie Village, KS 66208.

It is very exciting to learn to recognize the different species, to know which are rare treats to see, to become familiar with where they live, their habits, and what localities are relatively unspoiled by people. Learning about tiger beetles brings you closer to our natural world.

We are always interested in learning of new or unusual records of tiger beetles in our area. Please do not hesitate to write or e-mail us. We wish you many pleasant hours ahead.

Jonathan G. Leonard
106 Morrill Hall
University of Vermont
Burlington, VT 05405
jleonard@zoo.uvm.edu

Ross T. Bell
Biology Department
Marsh Life Science Building
University of Vermont
Burlington, VT 05405
rtbell@zoo.uvm.edu

THE AUTHORS

Jonathan Leonard is the computer teaching laboratory coordinator and lecturer for the College of Agriculture and Life Science at the University of Vermont. He is a recipient of the Kroepsch-Maurice Award for excellence in teaching at the University of Vermont, and teaches introductory computer applications. In recent years Jonathan has taught classes in World Food and Sustainable Development, Limits to Growth, and Natural History of Vermont.

Jonathan received his B.A. in Zoology from Drew University. At the University of Vermont, he earned a Master of Science in Zoology, and a Ph.D. in Plant and Soil Science, specializing in Entomology. He is the author of many entomological journal articles and scientific illustrations, including a cover illustration for the *Coleopterists' Bulletin*. Jonathan's research interests include social insects, periodical insects, alpine biology, sustainable human systems, and conservation biology.

In the 1970s and 1980s, Jonathan worked for the Appalachian Mountain Club, where he lived above treeline for months at a time. During this time he also worked in the Worcester Massachusetts Science Center Planetarium. An avid bicyclist, Janathan spent three months hitchhiking through Europe, across the Sahara Desert, and through West Africa. He has also ridden his bicycle from San Francisco to Alaska and from Vermont to Nova Scotia.

Jonathan plays several musical instruments, and one of his ambitions (however illusory) is to become the world's best bluegrass guitarist. He lives with his wife and daughter in Richmond, VT.

Ross T. Bell is the John Purple Howard Professor of Natural History at the University of Vermont, where he has taught for over 40 years. He was born in Urbana, IL, where he was educated. He received his Ph.D. degree in Entomology from the University of Illinois. Ross is a Research Associate with the Carnegie Museum of Natural History in Pittsburgh, PA.

The focus of Ross's scientific work has been the Carabidae, or ground beetles, on which he has published over 30 scientific articles, including large monographs. For the last 30 years, he and his wife and collaborator, Joyce R. Bell, have been working on a monograph of the rhysodine beetles, a group of highly specialized ground beetles inhabiting dead wood. This project has taken them to many European museums, as well as to countries with tropical and subtropical rain forests. Their adventures include a year in Australia and three months in Papua New Guinea studying beetles.

Ross is a member of numerous entomological societies, and he, his wife, and co-author Jonathan Leonard, are founding members of the Vermont Entomological Society.

ACKNOWLEDGMENTS

We wish to thank Phil Nothnagle for unwittingly tossing the intellectual pebble that started our avalanche of interest in tiger beetles, and Katherine Brown-Wing, whose artwork inspired us. Mary Parkin (U.S. Department of the Interior) kindly sent us copies of the U.S. Fish and Wildlife's Recovery Plans for *Cicindela puritana* and *Cicindela dorsalis dorsalis*. The following persons were generous in sharing locality information, their knowledge of the biology and ecology of our species, and permission for using figures from their research. All provided encouragement that kept the work moving forward: Bob Acciavatti (Carnegie Museum of Natural History), Yves Bousquet (Agriculture Canada), Howard P. Boyd (American Entomological Society), Paul Brunelle (Halifax, NS), Bob Davidson (Carnegie Museum of Natural History), Gary A. Dunn (Young Entomologists' Society), John R. Grehan (Pennsylvania State University), Lee H. Herman (American Museum of Natural History) Walter N. Johnson (Minneapolis, MN), Bruce R. Kirby (Smithsonian Institution Archives), C. Barry Knisley (Randolph-Macon College, VA), Andre Larochelle (Manaaki Whenua Landcare Research, New Zealand), David J. Larson (Memorial University, Newfoundland), David McCorquodale (University College of Cape Breton), Donald H. Miller (Lyndonville State College, VT), Robert Nelson (Colby College, ME), Philip Nothnagle (Windsor, VT), Raymond J. Pupedis (Peabody Museum, Yale University), Peter Rankin (University College of Cape Breton), Brendan Reardon (Winchester, MA), Howard Romack (Cambridge, NY), Tom Schultz (Denison University, OH), John Stamatov (Armonk, NY), and Harold L. Willis (Wisconsin Dells, WI). Thanks go to our wives, Denise Martin and Joyce Bell, for support during the writing of this book, and to Emma Leonard who helped discover *Cicindela marginipennis* on the Winooski River. We are grateful to Philip Nothnagle and Howard Boyd for their thoughtful review and suggestions for improving the manuscript. We are especially indebted to our editor, John Sulzycki and Project Editor, Helen Linna.

CONTENTS

GREEN SPECIES

DARK SPECIES WITH COMPLETE WHITE MACULATIONS

DARK SPECIES WITH MACULATION PATTERN REDUCED

BRONZED SPECIES WITH REDUCED MACULATIONS AND PROMINENT MIDDLE BANDS

SPECIES WITH DISTINCTIVE MARGINAL BANDS

SPECIES WITH PALE ELYTRA

STRAY SPECIES AND QUESTIONABLE RECORDS

INTRODUCTION

The beetles, or the great insect order Coleoptera, are extremely diverse and contain more species than any other order in the animal kingdom (see Glossary at the end of the book for unfamiliar terms). The observant naturalist soon discovers that beetles fall into family subgroups such as ground beetles (Carabidae), scarab beetles (Scarabaeidae), or weevils (Curculionidae). Current classification places the tiger beetles in the ancient ground-beetle family Carabidae (see Table 1 for an example of the current classification). In most insect and beetle books, the tiger beetles and the rest of the ground beetles are found near the beginning (e.g., White 1983).

Tiger beetles are among the many groups making up the larger family of ground beetles (Carabidae). Tiger beetles are clearly more closely related to some ground beetle tribes, such as the burrowing scaritines, than these tribes are to some other ground beetle tribes, such as the snail-eating cychrines. For this reason, ground beetle experts include tiger beetles in the family Carabidae as a subfamily (Cicindelinae) (e.g., Erwin 1979). Tiger beetle experts however, usually favor following an older tradition of making them a separate family (Cicindelidae), and justify it by pointing out the many structural features which have been changed in the evolution of tiger beetles from ground beetles (e.g., Crowson 1981).

HOW TO RECOGNIZE TIGER BEETLES

Tiger beetles differ from the rest of the ground beetles in having their antennae attached to their head dorsomedial (above and to the middle of the body) of the jaw articulation, while the rest of the ground beetles have the antennae attached to the head below the eye, between the eye and mandible (Fig. 2). The location of the antenna is the one most easily recognized characters for all tiger beetle species worldwide. Local tiger

Table 1:   Classification of the Northeastern Beach Tiger Beetle

Kingdom:	Animalia (Animals)
Phylum:	Arthropoda (Animals with jointed legs)
Class:	Insecta (Insects)
Order:	Coleoptera (Beetles)
Suborder:	Adephaga
Family:	Carabidae (Ground Beetles)
Subfamily:	Cicindelinae (Tiger Beetles)
Genus:	*Cicindela* (Diurnal Tiger Beetles)
Species:	*dorsalis* (North American Beach Tiger Beetle)
Subspecies:	*dorsalis* (The Northeastern Beach Tiger Beetle)

beetles have huge bulgy eyes (Figs. 3 and 4), but two genera in the western U.S. have small eyes (*Amblycheila* and *Omus*).

All tiger beetles known to be established in the northeast belong to the genus *Cicindela*. They are fast-moving, alert insects with flat elytra without striae or longitudinal grooves (Figs. 5 and 6). Most species have white marks or maculations on their elytra. When approached, tiger beetles fly several yards away, often landing to face the intruder.

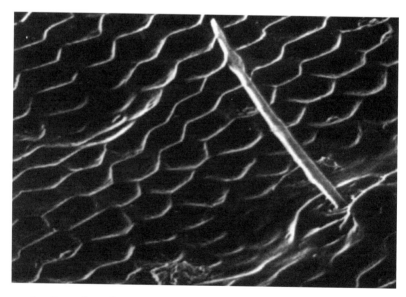

Figure 1. Scanning electron micrograph of the elytral surface of *Cicindela sexguttata* magnified 800 diameters. Hexagonal alveoli are seen on the surface along with a tiny seta on the right.

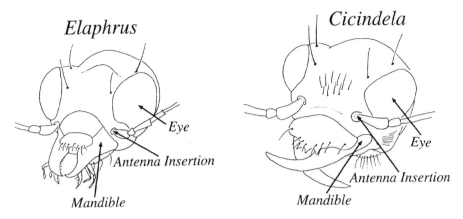

Elaphrus

Cicindela

Eye

Antenna Insertion

Mandible

Eye

Antenna Insertion

Mandible

Figure 2. Heads of the ground beetles *Elaphrus* and *Cicindela*. Note where the antennae attach to the head. In tiger beetles, the antenna insertion is between the eye and the labrum, far above the articulation of the mandible.

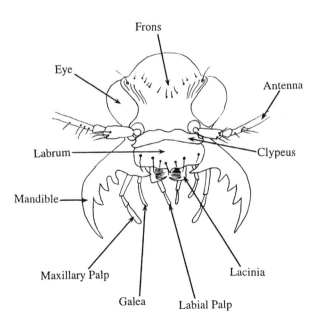

Frons

Eye

Antenna

Labrum

Clypeus

Mandible

Maxillary Palp

Lacinia

Galea

Labial Palp

Figure 3. Front view of the head of *Cicindela repanda,* showing the characters used in the identification keys.

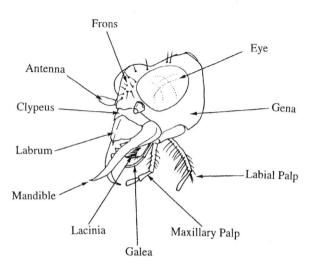

Frons

Eye

Antenna

Clypeus

Gena

Labrum

Labial Palp

Mandible

Lacinia

Maxillary Palp

Galea

Figure 4. Side view of the head of *Cicindela repanda,* showing the characters used in the identification keys.

Beetles that are similar to tiger beetles and could possibly be mistaken for them include some ground beetle species, especially the large-eyed species of *Elaphrus* (Fig. 7). These are active by day, and are usually found on bare sand or mud, where tiger beetles may also be present. Beginners often think *Elaphrus* are small tiger beetles. However, *Elaphrus* lack white marks on the elytra, and have large circular pits (foveae) with ring-like outer edges on the elytra. In addition, like other ground beetles in the narrow sense, they have their antennae arising from the sides of the head, between the jaw articulations and the eyes (Fig. 2).

Approximately 2000 species of tiger beetles are known and are found worldwide, except in Tasmania, Antarctica, and some remote oceanic islands (Kryzhanovskiy 1976; Pearson 1988; Pearson and Cassola 1992). In North America, 147 species have been described (Boyd 1982). In the area covered in this field guide, there are 20 species. Four of these species have two subspecies or races, for a total of 24 forms. Two additional species reported from our area are questionable or strays. One of these species is represented by a single individual (*Cicindela trifasciata ascendens*), evidently a stray from further south. The second species (*Tetracha carolina carolina*) is also a possible stray migrant or is simply a mislabeled specimen. Larochelle (1986b), Bousquet and Larochelle (1993) and Pearson et al. (1997) published three excellent sources of information on the geographical distribution of tiger beetles for all of North America north of Mexico.

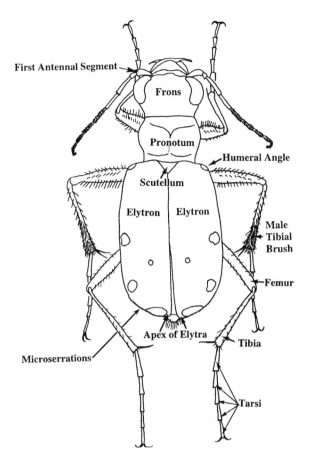

Figure 5. Dorsal view of *Cicindela sexguttata* male, showing the characters used in the identification keys.

WHAT IS A SPECIES?

The common everyday understanding is that species are "kinds" of animals that form unique populations. Different species are recognizable and separable from other species by morphological differences. This concept of species can be termed the "morphological species concept." For most biologists, the most important criteria defining different species is whether populations are interbreeding units. Two organisms, by this "biological species concept," belong to the same species if they can interbreed and produce viable fertile offspring. By this definition, genetic isolation defines the arena in which the evolution of different species takes place.

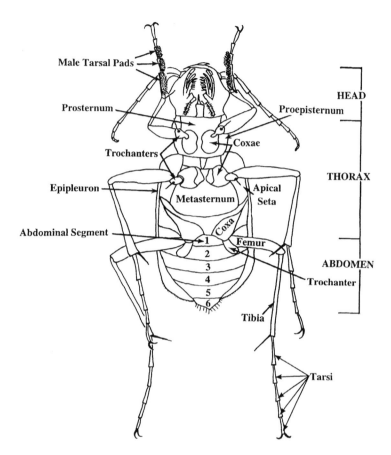

Figure 6. Ventral view of *Cicindela sexguttata* male, showing the characters used in the identification keys.

Often two species are isolated from each other by space (i.e., mountains, oceans, deserts), time (i.e., they emerge from pupae at different seasons), or by behavior. Sometimes two species can not be easily distinguished by morphology, but in fact they are not interbreeding. Species of this type are termed "cryptic" or "sibling" species.

Each species is given an italicized binomial or trinomial name followed by the author and date of publication of the species description. For example, *Cicindela dorsalis dorsalis* Say, 1817. The first name (starting with a capital letter) is the genus name. A genus is a group or clade of similar and evolutionarily related species. The second name (starting with a small letter) is the species. Note that the species name is not complete without the genus name. Just as "Jane" is not a complete personal name

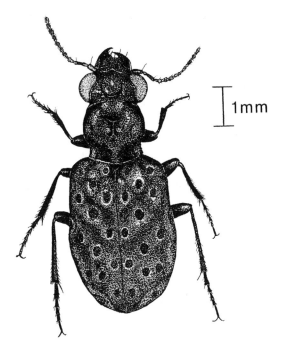

Figure 7. Habitus drawing of ground beetle *Elaphrus olivaceus* LeConte, 1863.

for a human individual, you need "Jane Smith" to get a library card or a driver's license. For informal usage, when it is obvious that *Cicindela* is the genus meant, we can abbreviate the genus name, as in *C. repanda*, just as "Jane S." might be clear enough if you know that a small group of people is being discussed. The third name, or trinomial, is the subspecies name.

A problem occurs when systematists try to decide if a species with two or more color or morphological forms is really more than one species. In this case systematists look through museum collections to try to find intergrades (individuals that have intermediate characteristics of the different groups). Information about the geographical areas in which the two forms occur is used as evidence to decide whether there are two species or one. If the two color or morphological forms do not overlap (allopatric populations), a judgement is made about the degree of difference between the two populations. If there is a clear distinction, the two forms will often be declared two separate species. When populations border one another (parapatric populations), evidence of interbreeding is sought in the area where they overlap. If no intergrades can be found from a large collection of individuals in an area of overlap, this suggests that the groups are two separate species. If there are intergrades, then the groups are assigned to the same species, with a trinomial used to identify geographically distinct subspecies. For example, *Cicindela dorsalis dorsalis* is found from Massachusetts to Virginia, and *Cicindela dorsalis media* is found from New Jersey to Florida. Intergrades can be found in New Jersey and Chesapeake Bay (Boyd and Rust 1982).

In the past, some entomologists have used the third name in the trinomial for "morphs" or "varieties" to represent extremes of color or morphological variation of a species within a single geographical area. This practice is now discouraged because of the confusion it causes. The third name in the trinomial should only be used for geographically distinct subspecies.

We have to make at least provisional decisions about status of populations in order to put them into the classification system, so we can refer to them in the future. Entomologists recognize that classifications are often best estimates that can be made under the circumstances, and we expect some of them to be changed when more information is known.

Species names are formally followed by the last name of the person who first described them and the date of the publication of the description. For example, *Cicindela rufiventris heutzii* Dejean, 1831 is the genus *Cicindela* (diurnal tiger beetles), species *rufiventris* (red-bellied) and subspecies *heutzii*, named by Dejean in 1831. To further complicate matters, the name was probably chosen to honor the eccentric entomologist Nicholas Hentz (see Hanley 1977 for a discussion of Hentz), but was published as "heutzii" instead of "hentzii," due to a typographical error. Unfortunately, the original published spelling must be accepted as the valid name according to the zoological rules of nomenclature. The reason the author and date are cited in the full species binomial or trinomial name is to reduce the likelihood that the same name might be used by more than one person for two different kinds of animals. If a later study presents convincing evidence that the species belongs in a different genus, the author's name and date are thereafter written in parenthesis; for example *Tetracha carolina carolina* (Linné 1767).

Insect species identification and classification is usually based upon morphological characters (number of hairs, coloration, and shape of body parts, etc.). Taxonomists and field naturalists use characters they can see with their eyes or with a hand lens or microscope. Most often, species differences are very obvious. However there are cases where color, size, and shape variation makes it difficult to assign an individual to one species or another. In this case, the taxonomist might label the specimen with only the genus name, as in "*Cicindela* sp."

The species reported from our area fall into two genera: *Tetracha* (one species) and *Cicindela* (the remaining 21 species). What we call *Tetracha* is referred to as *Megacephala* in recent literature (Huber 1994). In south and western North America are two other genera, *Omus*, that includes approximately 7 to 10 species of large black robust wingless tiger beetles that hunt at night, and *Amblycheila*, containing half a dozen species of truly gargantuan tiger beetles (body length up to 38 millimeters or 1.5 inches long!) (Leng 1902).

TIGER BEETLE ECOLOGY

In our area, adult tiger beetles are most often found on warm sunny days on open ground where they run rapidly across the soil, pause, then run again in search of prey. They will run towards, overpower, and eat small arthropods they detect with their large eyes. Ants are one of their favorite foods. The tiger beetle chews the prey into fragments that are formed into a ball in its mouth. Powerful enzymes dissolve the flesh of the prey. These digestive enzymes can also dissolve holes in a collecting net (Philip Nothnagle, personal communication). The tiger beetle swallows only the resulting "soup," and drops a ball of dry fragments when it is through (Evans 1965). The totally liquid diet means that you can not study the beetle's feeding preferences by identifying the insect scraps in its stomach contents, as can be done for many other predatory insects.

The larva (Figs. 8 and 9) typically excavates a vertical hole or burrow. It waits for prey at the top of the burrow, with the flat top of the head and the first body segment (prothorax) filling the opening. These exposed body parts are often brightly colored and can match the surrounding soil color. When an ant, spider, or other potential meal ventures too close, the larva pops out like a jack-in-the-box and seizes it. Only the front part of the larva comes out of the burrow. The fifth segment of the abdomen has a hump with projecting hooks (Fig. 10). This prevents the larva from being pulled out of the hole by the prey. It also holds the larva in position while it is waiting for prey. The pattern of setae on the back of the pronotum and the shape and number of setae on the median and inner hooks are used to distinguish species. If you approach slowly, you can sometimes see a larva waiting, with its head looking like a door to the burrow. If you frighten the larva, it releases its hold on the burrow wall and drops quickly to the bottom. In the tropics, larvae in the genus *Collyris* excavate a burrow in the pith of tree twigs (Hamilton 1925).

Many of the details of the lives of tiger beetles were first discovered by the great ecologist Victor E. Shelford, who made countless hours of

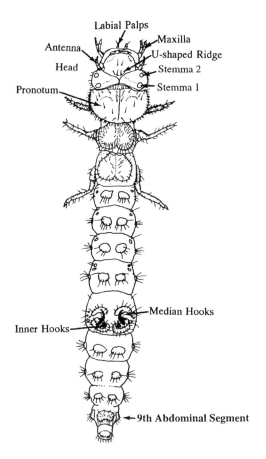

Figure 8. Dorsal view of *Cicindela purpurea* larva. (Redrawn with permission from Shelford 1908.)

observations in the field and in his laboratory at the University of Illinois. His published papers are classics (Shelford 1908, 1917) as is his *magnum opus* book, *The Ecology of North America* (1963).

Adult tiger beetles make shallow burrows in the soil for protection during the night and deeper burrows for overwintering (Leng 1902). Overwintering burrows (hibernacula) vary in length from 15 centimeters (6 inches) for *C. repanda* to 122 centimeters (48 inches) for *C. formosa* (Wallis 1961). Each species has its own preference for type of soil, slope, salinity, and vegetation. To make these burrows, the adult uses its mandibles to loosen the soil, pushes the soil into a pile under its body with its front legs, passes the pile back with the middle legs, and then uses the hind legs to kick the soil free of the burrow entrance. After the

overwintering burrow is partially excavated, the beetle stops kicking the soil out of the burrow entrance, and packs it in the upper tunnel of the burrow, sealing itself inside. When the burrow is complete, the beetle crawls to the bottom, turns around, and stays there all winter in hibernation.

The life cycle of *Cicindela purpurea* is typical of many tiger beetles in our area. During mating, the male rides on top of the female and holds on to the female's prothorax with his jaws and front legs (see tarsal pads on male, Fig. 6). A few days after mating and fertilization, the female lays about 50 eggs, each in its own tiny uncovered hole about 5 to 10 mm deep that is dug by the female's ovipositor. The eggs are cream colored and about 2 mm long and are shaped like elongated chicken eggs. The female chooses a particular soil texture, moisture, and exposure to lay her eggs, often probing the ground with her ovipositor at a few different locations before actually laying an egg. The larvae hatch from eggs after about a 2-week incubation period in summer, often after a rain (Knisley and Schultz 1997). The first larval stage or instar often lasts from 4 to 5 weeks; then the larva closes its tunnel, retreats to the bottom of the burrow, molts into the second instar stage, and opens the burrow again about a week later. All tiger beetle larvae go through three instars. The larva must widen and deepen the burrow to accommodate its larger size after molting. The second instar lasts about 5 to 7 weeks, then the larva molts again into a third instar larva. If the soil becomes unsuitable, the larva will abandon the burrow and seek out better soil. Most often larvae over winter as third instars. Before pupation, the third-instar larva excavates a chamber to the side of the main burrow, where it pupates for 2 to 4 weeks (Figs. 11 to 12). More information on the ecology of larvae can be found in the papers by Shelford (1908) and Hamilton (1925) and in Knisley and Schultz's (1997) book.

The presence and correct identification of larvae are very important for tiger beetle conservation, because larvae show that a reproducing population is locally established. In many species, adults can disperse great distances, but then may die without reproducing. Therefore, the presence of larvae means a viable population is present.

The burrows of tiger beetles are very distinctive, and you can often identify the species just from the shape of the burrow. Some burrows look as if someone drilled a clean hole straight down into the soil. Other burrows enter the soil at an angle leaving an oval entrance. Some burrows have cone-shaped funnels leading down to the beginning of the burrow tunnel, while still others have a pit just below the entrance, and the burrow tunnel descends unseen from one side of the underground pit. Burrow entrances of most species are very smooth and clean with the soil tamped down immediately around the entrance; this feature usually

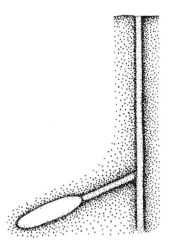

Figure 11. **Burrow of *Cicindela scutellaris* showing side pupal chamber.
(Redrawn with permission from Shelford 1908.)**

distinguishes tiger beetle burrows from holes dug by worms or other
invertebrates. Also, tiger beetles will clear out their burrows after a rain
and toss soil pellets out of their burrow, creating a small pile about 5 to
15 cm away from the burrow entrance (Knisley and Schultz 1997). Some
spiders make very similar looking burrows. The spider burrow, however,
is lined with silk, as will become evident if you probe just inside the
entrance with a thin grass stalk. To see if a larval burrow is occupied,

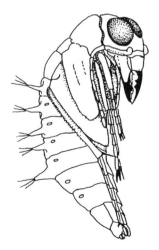

Figure 12. **Pupa of *Cicindela purpurea*. (Redrawn with permission from Shelford
1908.)**

put a long straight grass stalk down as far as it will drop. Wait a few minutes. If a larva is present, it will push at and try to remove the stalk.

The life cycles of tiger beetles fall into two general categories: "spring–fall" and "summer." Those with the spring–fall life cycle emerge from their pupal chambers as adults in the late summer or fall, may be active for several weeks, then over winter in burrows. The following spring the now sexually mature adults emerge from hibernation, mate, lay eggs, and die. Larvae may live a year or more before pupating in the summer to emerge as adults in the fall. This spring–fall life cycle can take 1 to 4 years to complete, depending on food supply, temperature, and length of the growing season. The life cycle of tiger beetles reared in the laboratory has been artificially extended or reduced by a year by feeding them more or less food, respectively. Adult emergence often occurs after a soaking rain (Knisley and Schultz 1997). Most of the species in our area have this spring–fall life cycle and complete it in 2 years. Many of these species have more than one "brood" present at any one location. For example, in late summer while newly emerged adults are getting ready to overwinter, the third instar larvae of the next brood are also present.

The "summer" type of life cycle can be 1 or 2 years long (possibly longer in higher latitudes), and begins when adults emerge in early summer. They mate and lay eggs in midsummer; then the adults die in late summer. The larvae grow, molt, and hibernate as second or third instars. Some individuals or populations take another year before pupating in spring and emerging as adults in early summer. The northeastern beach tiger beetle (*Cicindela dorsalis dorsalis*) and the common punctured tiger beetle (*C. punctulata*) both have a summer life cycle.

The adult tiger beetle changes color as it ages. Beetles are very pale and soft-bodied during the first 48 hours after emergence from the pupa. Such a specimen is termed "teneral." Tiger beetles remain in their pupal chamber during this time while the color pigment develops. Recent studies by Tom Schultz and his students have shown that during this period the melanin layers in the cuticle become thicker, causing a color change from shorter to longer wavelengths (Knisley and Schultz 1997). The body color of adults usually darkens over the following weeks as the beetle becomes older. In the laboratory, Victor Shelford reared larvae under different conditions of temperature and moisture, resulting in variations of adult maculation pattern, color shade, and hue (Shelford 1917).

Male tiger beetles can be distinguished from females by dense brush-like pads on the first three tarsal segments of the forelegs (Fig. 6). Also, males have pubescent hairs on the outer sides of the middle tibiae (Fig.

5). Females have slender front tarsi with smooth middle tibiae. Females also have a shallow groove or fold on the sides of their prothorax, the proepisternum (Fig. 6), that males grip with their mandibles while mating (Freitag 1974). This groove is absent in males. In some species there are further differences, in the shape of the elytron, the color of the labrum, the teeth on the mandible, or the distribution of hairs or setae on the face.

Population dynamics of tiger beetles are affected by abundance of food, suitable soil for burrowing, weather, and the presence of natural enemies including predators and parasites (Pearson 1988; Knisley and Schultz 1997). Predators of tiger beetle adults include spiders, robber flies (Asilidae), toads, birds, and mammals such as raccoon and skunk (tiger beetle elytra have been found in scat). Predators of tiger beetle larvae have not been studied extensively, but include ants, hister beetles (Histeridae), soldier beetles (Cantharidae), and birds (Knisley and Schultz 1997).

Parasitoids of tiger beetle larvae include Bombyliid flies (Diptera: Bombyliidae) of the genus *Anthrax*, mites, and tiphiid wasps (Hymenoptera: Tiphiidae) of the genera *Methoca* and *Pterombrus*. In some populations of the smooth tiger beetle, *Cicindela scutellaris*, over 80% of the larvae were attacked by *Anthrax*, although usually rates were much lower (Shelford 1913). Parasitism of tiger beetle larvae due to tiphiid wasps have been reported at rates upwards to 63% (Knisley and Schultz 1997).

HOW TO RECOGNIZE THE DIFFERENT SPECIES

In the field, you can recognize many of our species without getting very close to the beetles or larvae. Close-focusing binoculars are very useful in identifying tiger beetle species in the field. Often the adult elytral maculations are distinct enough to identify the species. The nomenclature used for elytral markings is shown in Fig. 13. The time of year, geographical location, and habitat can greatly increase the probability of certain species being present. Look for adult beetles on sunny days after a soaking rain. Use the life cycle, habitat, and range information in the species descriptions further on in this book to determine what species are likely to be found where and when. The key characters you can use in the field to identify the species are underlined in the species descriptions.

The beginning cicindelaphile (tiger beetle lover) will need to catch most tiger beetles and view them up close for species identification. After some time collecting, you will be able to identify beetles to species at a glance from behavioral cues along with habitat and time of year information.

An insect net is a must to catch almost all tiger beetles. Nets may be purchased from various biological supply houses, such as:

Carolina Biological Supply (2700 York Road, Burlington, NC 27215, Phone: 800-334-5551)

Connecticut Valley Biological (82 Valley Road, P.O. Box 326, Southhampton, MA 01073 Phone: 413-527-4030)

BioQuip Products (17803 LaSalle Ave. Gardena, CA 90248-3602, Phone: 310-324-0620)

Wards Natural Science Establishment, P.O. Box 92912, Rochester, NY 14692-9012 Phone: 800-962-2660)

It takes practice to net tiger beetles. One must slowly sneak up to the beetle, keeping low to the ground. Beetles will almost always take off just as you are starting to swing your net so you must either slap the net to the ground with the tiger beetle inside, or anticipate where the beetle will fly and swing your net to intercept it along its flight path. More often than not, the beetle flies in a direction not anticipated. In loose sand, a

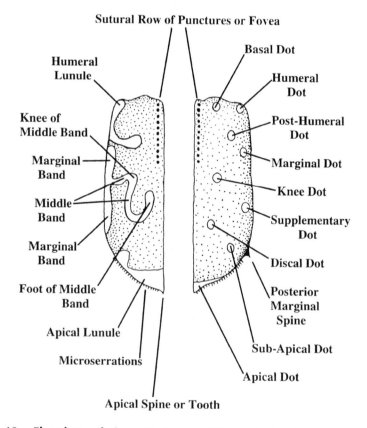

Figure 13. Elytral maculations. (Redrawn with permission from Boyd 1978.)

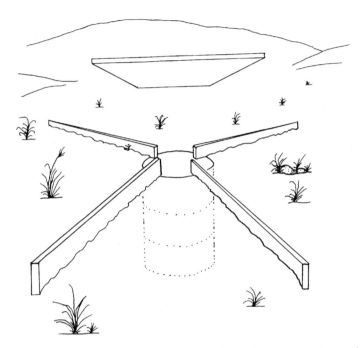

Figure 14. Pitfall trap using coffee can and baffles with a board for a rain cover.

tiger beetle will often burrow in and tunnel its way out from under the net before you can get hold of it. Patience and practice will be rewarded.

Another method of catching tiger beetles is with pitfall traps. Use large empty coffee cans, or similar-sized containers, and bury them with the rim of the can at soil level (Fig. 14). The can needs to be deep enough so the beetles cannot jump out (a cup is not deep enough). Pitfall trap efficiency can be improved by using at least three baffles that channel the beetles toward the lip of the trap. A board of wood or similar cover keeps the rain out if traps are left for extended periods of time.

Once beetles are caught, a magnifying glass or hand lens is often necessary to find the characters to identify the species. The hand lens should magnify between 10 and 20 times. Glass vials with screw tops are useful to contain the beetles as you are observing them with the hand lens (adults will bite if handled, but the bites are not extremely painful). A camera with macro lens and flash is especially useful to record individuals that you will subsequently release. We recommend a macro lens with a focal length of at least 105 mm, so you do not have to bring the camera right next to the specimen to get a good photograph. The camera and glass collecting vials are a must to record new localities for endangered or threatened species, since it is against federal or state law to collect (and pin) these species. It is important to record the date, geographical location, and habitat information of each species photographed (we either write this on the back of the photo, or mount the photo in our field journals).

Most entomologists studying tiger beetles have a reference collection of pinned specimens. There are arguments for and against keeping a collection. Although unlikely, it is theoretically possible to over-collect at a locality and cause local extermination of a species. However, there is very little evidence that this has ever happened with insects. In fact, in many cases, attempts to collect an entire local insect population led to the immigration of more individuals from other nearby populations (Knisley and Schultz 1997). Most entomologists would argue that if one does not collect every individual seen, the advantages of collecting and having pinned specimens as a record of where the species was found far outweighs any possible danger of causing a local population extinction. Most important for the survival of the tiger beetle population is preservation of its preferred habitat and food source. Habitat destruction and degradation by "all terrain" vehicles such as jeeps and dune buggies is a much more serious threat to the survival of tiger beetle populations than is collecting. Prudent collectors will not attempt to take every specimen they see, but will collect only a few individuals. Also some individuals may be photographed and then released. The pinned specimen has the advantage that it can be preserved indefinitely, can be observed again and again (possibly identified as a different species than was originally thought), and contains more information than a photograph. However, endangered or threatened species should not be collected. In our area, the puritan tiger beetle (*Cicindela puritana*), the northeastern beach tiger beetle (*Cicindela dorsalis dorsalis*), and the cobblestone tiger beetle (*Cicindela marginipennis*) should **never** be collected. In Vermont, the hairy-necked tiger beetle (*Cicindela hirticollis*) is threatened, as is the red-bellied tiger beetle (*Cicindela rufiventris*) in Connecticut. These species should not be collected. Since it is a crime to collect or harass these species, they should be photographed instead.

Insects preserved in a collection should be pinned and stored in tight insect drawers or boxes with fumigant. We use Cornell drawers made from kits ordered from a biological supply house. Beetles freshly caught should be killed using an agent such as ethyl acetate, pinned, and legs arranged on a Styrofoam block. A collection label must accompany each specimen with state, county, town, date, and collector information. Ideally, specific habitat and elevation information should be included. (See following sample for actual size and enlarged to show information.)

VT: Chittenden Co., Bolton
Gravel pit, elev. 250 m
15 May 1995
J. G. Leonard, Coll.

Labels should be printed on high-quality paper. We use 100% rag, acid-free thesis paper, available at college and university bookstores.

After the specimen has dried (usually a week at least), the collection label is placed on the pin underneath the insect, and the pinned specimen is transferred into the Cornell drawer or insect box. Often, dried specimens become greasy and dull. You can de-grease dried beetles by soaking them in ethyl acetate for several weeks (Knisley and Schultz 1997). Also, dropping dried beetles into boiling water for a few minutes will soften them and allow you to re-position their legs and antennae before re-drying, if necessary. Refer to a general entomology textbook or insect field guides for more information on making an insect collection (e.g., Martin 1977; White 1983).

The keys that follow can be used to identify adults and most larvae in our area to species. A key to adults for all North American species has been published by Harold Willis (1968) and maps showing the general distribution of most North American species have been published by David Pearson et al. (1997). Clyde Hamilton (1925) published a key to many species of larvae. A summary of North American larvae that have been taxonomically described is found in Valenti (1996a). Although under good illumination a 10 to 20× hand lens can be used to view most adult and larval key characters, a binocular dissecting microscope is usually preferred.

Collecting larvae is an important, but labor-intensive, job. The most common method is to carefully insert a long grass stem into the larval burrow, and then carefully dig a deep wide hole parallel to the burrow. Then the soil between the hole and the burrow is carefully removed from the top down until one reaches the larva, which is usually found at the bottom. Another method is the "sneak and stab" method described by Knisley and Schultz (1997). In this method, one sneaks up to a larva and uses a trowel to quickly stab the soil at a 45° angle, thereby trapping the larvae above the trowel. This method takes great patience and stealth, as the larvae, especially during the day, are very wary and will drop to the bottom of their burrows at any sign of danger. One way to get the larva near the entrance of its burrow is to insert a grass stem into the burrow. Often, the larvae will lift it up and try to remove it. As it is near the entrance the collector can stab the trowel. Barry Knisley and James Hill extended their reach by tying trowels to 1.5 meter-long bamboo poles. They then could stab and trap the larvae without getting very close to the burrow and alerting the larvae. They were very successful at collecting *Cicindela dorsalis dorsalis* larvae at night, using this method with headlamps or flashlights to locate active larval burrows (Clancy 1996; Knisley and Schultz 1997).

Another method of collecting larvae is "fishing," where one inserts a grass stalk into the larval burrow. If a larva grabs the grass stem it is sometimes possible to quickly jerk the larva out of the burrow. The larva may be able to get a firm grip on the grass stem if the end that is inserted is frayed (Knisley and Schultz 1997).

KEY TO ADULT TIGER BEETLES

1   Third palpomere on maxilla longer than fourth (last) palpomere (Fig. 15). Dorso-anterior edge of pronotum much wider than posterior edge (Fig. 16). Elytra covered with many obvious punctures. Large green beetles > 20 mm in length, some species with light yellow-tan apical markings.............................**Genus *Tetracha*** 2

 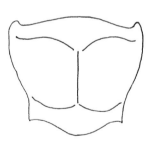

Figure 15. *Tetracha* maxillary palpus with 3rd and 4th palpomeres numbered.

Figure 16. *Tetracha* pronotum.

Third palpomere of maxilla shorter than fourth (last) palpomere (Fig. 17). Dorso-anterior edge of pronotum approximately same width as posterior edge. Base of elytra either without punctures, or if present found only along midline from the scutellum toward wing apex. Smaller beetles < 20 mm in length ....
....**Genus *Cicindela*** 3

Figure 17. *Cicindela* **maxillary palpus with 3rd and 4th palpomeres numbered.**

2   Large dark metallic-green tiger beetle with blackish color in midline. No markings on elytra ....
............................***Tetracha virginica*** (not described in this book)
Large metallic-green tiger beetle with red-bronze color in middle of elytra and two obvious light yellow-tan apical lunules ....
.... ***Tetracha carolina carolina*** (Plate 4)

3(1)   Beetles with elytra almost completely white 4
Beetles with elytra not completely white, but instead brown, black, green, or red, with or without white maculation patterns 5

4(3)   Body length >12.5 mm, with femur of hind leg very long and extending more than one third its length beyond body (Fig. 18).
.... ***Cicindela dorsalis dorsalis*** (Plate 4)

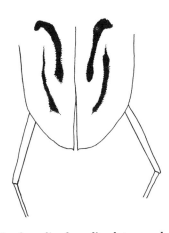

Figure 18. *Cicindela dorsalis dorsalis* **elytra and femurs. Note femurs extend far beyond elytra.**

Body length < 12.5 mm, with femur of hind leg not extending one third its length beyond body (Fig. 19) ....
....*Cicindela lepida* (Plate 4)

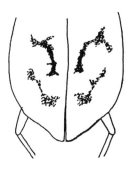

Figure 19. *Cicindela lepida* **elytra and femurs. Note femurs do not extend far beyond elytra.**

5(3) Length of labrum more than half the width ....
.... **Cicindela longilabris** (Plate 3)
Length of labrum less than half the width................................... 6

6(5) Ventral side of abdomen red or orange 7
Ventral side of abdomen metallic... 8

7(6) Elytral microserrations absent (Fig. 20). Maculations on elytra forming a band along the margin from humeral angle to apex. Middle
band missing. No dots on elytra ....
.... **Cicindela marginipennis** (Plate 4)

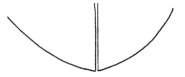

Figure 20. Elytra without microserrations.

Elytral microserrations present (Fig. 21). Maculations on elytra
include a middle band and lunules, or lunules reduced to dots
(usually 6 per elytron) and middle band extremely thin................
.... **Cicindela rufiventris** (Plate 2)

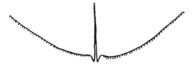

Figure 21. Elytra with microserrations.

8(6) Male with tooth on ventral side of mandible (Fig. 22). Female with strong elytral declivity at the apex near the mid-line (Figs. 23 and 96)..*Cicindela marginata* (Plate 2)

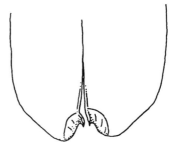

Figure 22. Tooth on ventral edge of mandible. (Redrawn with permission from Willis 1968.)

Figure 23. Deflexed elytra at the midline of *Cicindela marginata*.

Mandible without ventral teeth, female without strong elytral declivity 9

9(8) Labrum with single tooth (Fig. 24).. 10

Figure 24. Labrum with single tooth. (Redrawn with permission from Willis 1968.)

Labrum with 3 teeth (Fig. 25)... 18

Figure 25. Labrum with 3 teeth. (Redrawn with permission from Willis 1968.)

10(9)   Genae (Fig. 4) with setae...11
        Genae glabrous 15

11(10) Numerous white decumbent setae on the clypeus (Fig. 26)..........
        ....*Cicindela puritana* (Plate 2)

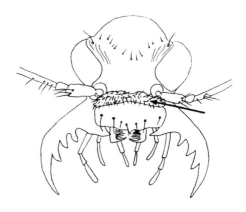

Figure 26. Clypeus with decumbent setae.

        Zero or possibly 1 or 2 setae on the clypeus (Fig. 27)..............12

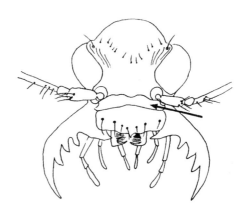

Figure 27. Clypeus glabrous.

12(11) Strong wide elytral maculations merging together. Humeral lunule
        connected to or almost touching the middle band.........................
        ....*Cicindela limbata labradorensis* (Plate 4)
        Elytral maculations narrow, not merging together. Humeral lunule
        not connected to or touching the middle band.........................13

13(12) Epipleura (Fig. 6) pale for most of their length, pronotum about as long as wide (Fig. 28)1 4

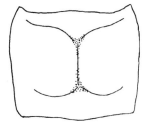

Figure 28. Pronotum narrow.

Epipleura dark for most their length, pronotum wider than long (Fig. 29)...............................***Cicindela duodecimguttata*** (Plate 2)

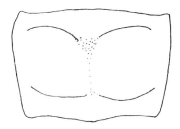

Figure 29. Pronotum wide.

14(12) Elytral maculations complete ....
.... ***Cicindela repanda repanda*** (Plate 1)
Elytral maculations reduced ....
.***Cicindela repanda novascotiae*** (not illustrated; see page 101)

15(10) Scape of the antenna with 1 seta (Fig. 30) 16

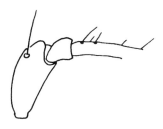

Figure 30. First antennal segment (scape) with only 1 sensory seta.

Scape of the antenna with 2 to 4 setae (Fig. 31) 17

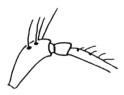

Figure 31. First antennal segment (scape) with 3 to 4 sensory setae.

16(15) Labrum (Figs. 3 and 4) with less than 8 setae. Elytral maculations reduced to tiny dots or absent. Green punctures near the midline...***Cicindela punctulata*** (Plate 3) Labrum with 8 or more setae. Elytral maculations complete with middle band long and S-shaped. Green spots may be present, but not limited to the midline .... ***Cicindela trifasciata ascendens*** (Plate 4)

17(15) Elytral maculations strong and complete... ....***Cicindela hirticollis hirticollis*** (Plate 2) Elytral maculations reduced ....***Cicindela hirticollis rhodensis*** (Plate 2)

18(9) Middle band absent .... 19 Middle band present.. 21

19(18) Body color bright metallic blue-green. Strong microserrations present (Fig. 21). Maculations reduced to dots.............................. .... ***Cicindela sexguttata*** (Plate 1) Body color reddish bronze to dull green. Microserrations are weak or absent. Marginal band is a prominent triangular mark 20

20(19) Elytra colored red-purple to bronze-brown..................................... .... ***Cicindela scutellaris lecontei*** (Plate 3) Elytra colored green to black .... ***Cicindela scutellaris rugifrons*** (not illustrated; see page 50)

21(18) Humeral lunule complete and intact... 22 Humeral lunule incomplete or absent 24

22(21) Marginal band connected to the humeral and apical lunules .... ***Cicindela formosa generosa*** (Plate 1) Marginal band not connected to humeral and apical lunules........ 23

23(22) Humeral lunule with long oblique posterior end................................
.... ***Cicindela tranquebarica*** (Plate 1)
Humeral lunule with posterior end abbreviated and slightly oblique,
resembling a short number 7 with a thick base.............................
.... ***Cicindela ancocisconensis*** (Plate 1)

24(21) Margin of elytra the same color as the dorsum...........................
.... ***Cicindela patruela*** (Plate 1)
Margin of elytra with dark indigo shadow or band contrasting with
the color of the dorsum of the elytra 25

25(24) Middle band forming a distinct knee and foot (Fig. 13). Posthumeral
and subapical dots large and obvious ....
....***Cicindela limbalis*** (Plate 3)
Middle band without obvious knee and foot. Posthumeral dot
absent or reduced to a tiny speck. Subapical dot often reduced
....***Cicindela purpurea*** (Plate 3)

KEY TO THIRD-INSTAR LARVAE

(Modified from Hamilton 1925)

1 Median hooks of fifth abdominal segment long, curved and sickle-
shaped, pointing outward (Fig. 32). Inner hooks short, cylindrical,
and usually with distal end suddenly constricted into a spine-like
projection. Ridge on caudal part of frons U-shaped and not joining
ridge on caudal part of the vertex (Fig. 33). Palpiger with a distinct
sclerite, proximal segment of labial palpus with 2 or 3 spine-like
projections on ventro-distal margin (Fig. 34) ....
....2. **Genus** ***Cicindela***

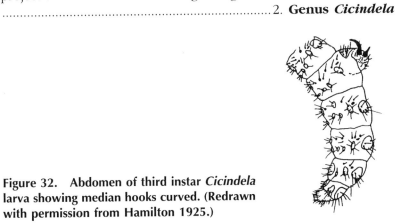

**Figure 32. Abdomen of third instar *Cicindela*
larva showing median hooks curved. (Redrawn
with permission from Hamilton 1925.)**

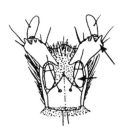

Figure 33. Ridge on frons not joining vertex ridge. (Redrawn with permission from Hamilton 1925.)

Figure 34. *Cicindela* palpiger with sclerite, and two to three spine-like projections on the labial palpus. (Redrawn with permission from Hamilton 1925.)

Median hooks thorn-like, straight or very slightly curved inward (Fig. 35). Inner hooks small versions of median hooks (about half the size). Ridge on caudal part of frons transverse and joining ridge on caudal part of vertex (Fig. 36). Palpiger membranous, proximal segment of labial palpus without spine-like projections on ventro-distal margin (Fig. 37).......................*Tetracha carolina* (p. 152)

Figure 35. Abdomen of third instar *Tetracha* larva, showing median hooks thorn-like and straight. (Redrawn with permission from Hamilton 1925.)

Figure 36. Ridge on frons joining vertex ridge. (Redrawn with permission from Hamilton 1925.)

Figure 37. *Tetracha* palpiger membranous, with no spine-like projections on the labial palpus. (Redrawn with permission from Hamilton 1925.)

2(1)   Maxillary palpus with 3 segments 3
       Maxillary palpus with 2 segments .....*Cicindela dorsalis* (p. 140)

3(2)   U-shaped ridge on caudal part of frons bearing 2 distinct setae
       (Fig. 38)..4

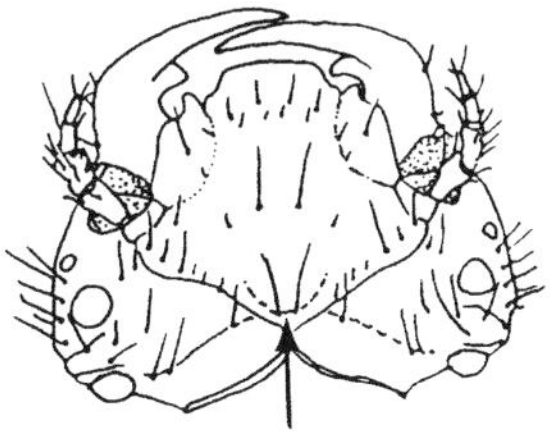

Figure 38. Two setae on U-shaped ridge on caudal part of frons. (Redrawn with permission from Hamilton 1925.)

U-shaped ridge on caudal part of frons bearing 3 or 4 distinct setae.. 17

4(3)   Inner hook with only 2 to 3 setae on shoulder (Fig. 39) 5

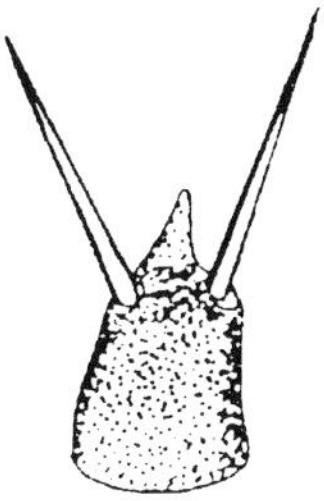

Figure 39. Inner hook with 2 setae on the shoulder. (Redrawn with permission from Hamilton 1925.)

Inner hook with more than 3 setae on shoulder (Fig. 40)............
.... *Cicindela marginata* (p. 84)

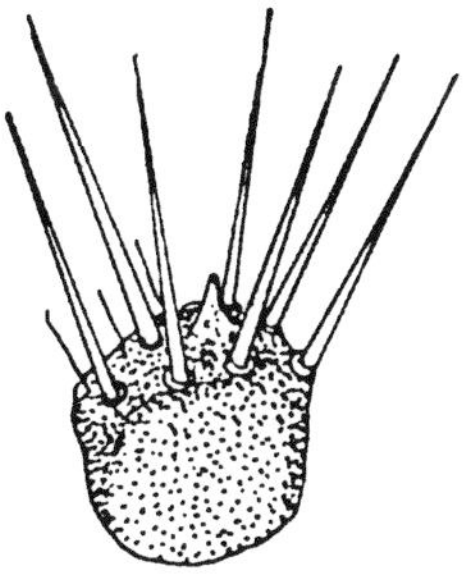

Figure 40. Inner hook with more than 3 setae. (Redrawn with permission from Hamilton 1925.)

5(4)   Pronotum entirely brown color with no light pattern. Dorsal hairs
       of head and pronotum brown........*Cicindela sexguttata* (p. 40)

Pronotum either metallic green, blue, bronzed brown, black or purple; if brown, a light pattern is obvious (Fig. 41). Dorsal hairs of head and pronotum white or transparent.................................. 6

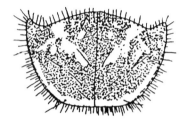

Figure 41. Light pattern on otherwise brown pronotum of the third instar larva of *Cicindela formosa generosa*. (Redrawn with permission from Hamilton 1925.)

6(5) Median hooks with 1 or 2 setae (or if 3, the 3rd seta is much smaller than other 2) 7

Median hooks with 3 distinct setae of similar size 12

7(6) Inner hooks with central spine projecting one third or more the entire length of hook (Fig. 42) 8

Figure 42. Inner hook with the central spine projecting one third or more the length of the hook. (Redrawn with permission from Hamilton 1925.)

Inner hooks with central spine projecting one sixth or less the entire length of hook (Fig. 43) 11

Figure 43. Inner hook with the central spine projecting one sixth or less the length of the hook. (Redrawn with permission from Hamilton 1925.)

8(7) Central spine of inner hook projecting more than one half the entire length of hook (Fig. 44) .... ***Cicindela limbalis*** (p. 118)

Figure 44. Central spine projecting more than one half the length of the hook. (Redrawn with permission from Hamilton 1925.)

Central spine of inner hook projecting one-half or less the entire length of hook (Fig. 45) 9

Figure 45. Central spine projecting one half or less the length of the hook. (Redrawn with permission from Hamilton 1925.)

9(8)   Pronotum with 4 to 6 pairs of secondary setae (Fig. 46) 10

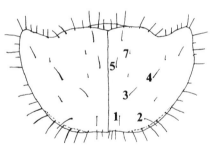

Figure 46. Pronotum of *Cicindela rufiventris* with 4 pairs of secondary setae. Primary seta number 6 is missing, but would be just lateral to 5 if it were present (primary setae are numbered, while secondary setae are unnumbered). (Redrawn with permission from Beatty and Knisley 1982.)

Pronotum with secondary setae missing except a single large seta Fig. 47, arrow) cephalo-lateral of seta 4 ....
.... *Cicindela purpurea* (p. 123)

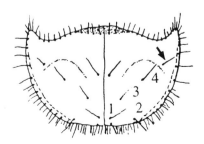

Figure 47. Pronotum with secondary setae missing except for one (arrow) cephalo-lateral of seta 4. (Redrawn with permission from Hamilton 1925.)

10(9)   Abdominal sclerites distinct. Diameter of stemma 2 greater than distance between stemmata 1 and 2 (See Fig. 8 for stemma numbering). Spine of inner hook one half the length of entire hook (Fig. 48). Head and pronotum black-brown color with green, purple, bronze, or copper reflections .... ***Cicindela rufiventris*** (p. 102)

Figure 48. Spine of inner hook one half the length of the entire hook. (Redrawn with permission from Hamilton 1925.)

Abdominal sclerites not distinct. Diameter of stemma 1 and 2 equal and equal to the distance between the two stemmata. Spine of inner hook one third the length of entire hook. Head and clypeus rufous with strong brassy and bluish-green metallic reflections; pronotum rufous with strong cupreous, violet, brassy, and bluish-green reflections .... ***Cicindela longilabris*** (p. 106)

11(7)   Pronotum U-shaped with less than 10 secondary setae (Fig. 49). Antenna with proximal segment bearing 10 to 11 setae. Median hooks with single large seta (or if 2, one is very small) .... ***Cicindela repanda*** (p. 52)

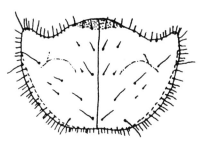

Figure 49. Pronotum U-shaped and with less than 10 secondary setae. (Redrawn with permission from Hamilton 1925.)

Pronotum shaped like a "V" with its bottom cut off, and covered with 50 or more secondary setae (Fig. 50). Antenna with proximal segment bearing 6 or 7 setae. Median hooks with 2 setae .... ***Cicindela lepida*** (p. 145)

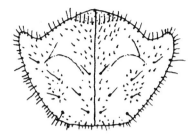

Figure 50. Pronotum V-shaped with a flat base and 50 or more secondary setae. (Redrawn with permission from Hamilton 1925.)

12(6) Spine of inner hook minute to less than one third the length of the hook.. ***Cicindela puritana*** (p. 73)
Spine of inner hook one third the length of the hook or
larger 13

13(12) Abdominal segment 3 bearing a moderate-sized sclerite with many setae anterior to the 2 coxal lobes.....***Cicindela patruela*** (p. 45)
Abdominal segment 3 bearing a small sclerite with only 1 or 2 setae anteriad of the 2 coxal lobes................................ 14

14(13) Ninth abdominal sternum with caudal margin bearing two groups of 4 setae each. Head and pronotum bright coppery bronze with a strong bluish-green reflection. Distance between stemmata 1 and 2 less than diameter of stemma 2..
.... ***Cicindela tranquebarica*** (p. 68)
Ninth abdominal sternum with caudal margin bearing two groups of 3 setae each (a 4th seta may be small)................................ 15

15(14) Antenna with proximal segment bearing 5 or more setae 16
Antenna with proximal segment bearing 4 setae ....
....***Cicindela ancocisconensis*** (p. 62)

16(15) Antenna with proximal segment bearing 8 to 11 setae. Pronotum with not more than 10 secondary setae and without any secondary setae on the meson between primary setae 1 and 5 (Fig. 51)......
.... ***Cicindela duodecimguttata*** (p. 96)

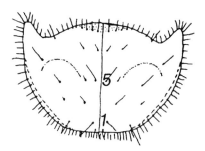

Figure 51. Pronotum of *Cicindela duodecimguttata* with less than 10 secondary setae and without a pair of secondary setae on each side of the meson between primary setae 1 and 5. (Redrawn with permission from Hamilton 1925.)

Antenna with proximal segment bearing 5 or 6 setae. Head and pronotum bronze with slight blue reflections. Pronotum with a pair of secondary setae on each side of the meson between primary setae 1 and 5 (Fig. 52). Abdominal sclerites not distinct. Diameter of stemma 2 less than distance between stemmata 1 and 2..........
....*Cicindela punctulata* (p. 111)

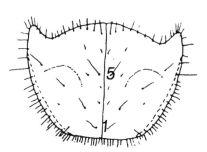

Figure 52. Pronotum of *Cicindela punctulata* with pair of secondary setae on each side of the meson between primary setae 1 and 5. (Redrawn with permission from Hamilton 1925.)

17(3)   Proximal segment of galea with 4 stout setae on the mesal margin. Diameter of stemma 2 distinctly less than distance between stemmata 1 and 2. Head and pronotum brown color with paler marks as shown in Fig. 53. Median hooks with 3 distinct setae. Average width of head and pronotum third instar 4.0 mm ....
....*Cicindela formosa generosa* (p. 57)

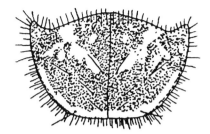

Figure 53. Pronotum of *Cicindela formosa generosa* showing pale marks on otherwise brown background. (Redrawn with permission from Hamilton 1925.)

Proximal segment of galea with 3 stout setae on mesal margin (Fig. 54). Diameter of stemma 2 similar to or greater than distance between stemmata 1 and 2. Head and pronotum not brown but metallic color 18

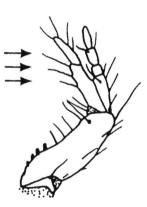

Figure 54. Proximal segment of the galea with 3 stout setae (arrows.) (Redrawn with permission from Hamilton 1925.)

18(17) Inner hooks with 4 or rarely 3 setae (Fig. 55). Pronotum with less than 50 small, not flattened secondary setae .... ***Cicindela scutellaris lecontei*** (p. 130)

Figure 55. Inner hook of *Cicindela scutellaris lecontei* with 3 to 4 setae. (Redrawn with permission from Hamilton 1925.)

Inner hooks with 2 setae (Fig. 56). Pronotum with 100 or more short, flat secondary setae (Fig. 57). Median hooks with 2 or 3 setae... 19

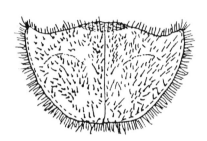

Figure 56. Inner hook with 2 setae. (Redrawn with permission from Hamilton 1925.)

Figure 57. Pronotum of *Cicindela hirticollis* larva with more than 100 short, flat secondary setae. (Redrawn with permission from Hamilton 1925.)

19(18) Median hooks with two setae. Inner hook spine one quarter the length of the entire hook. About 200 very flattened sac-like secondary setae on pronotum................***Cicindela hirticollis*** (p. 79)
Median hooks with 3 setae. Inner hook spine one third the length of the entire hook. About 100 slightly flattened secondary setae on pronotum..***Cicindela limbata*** (p. 88)

REARING TIGER BEETLES

Victor Shelford (1908: 159) described rearing tiger beetles as "very difficult" but went on to say that he had good success with a well-ventilated terrarium filled with the soil in which female beetles prefer to lay eggs. The top of the terrarium must have a metal flange projecting a few inches in from the edge to prevent the larvae from crawling up the side and escaping from the terrarium. Shelford had the best success with a terrarium with a fine wire-mesh bottom. The whole terrarium was set on moist or dry soil which thereby controlled the moisture of the terrarium soil. Shelford also made very narrow rearing chambers for observing the larvae. These were similar to "ant farms," where two plates of glass were separated by the same distance as the width of the tiger beetle larva's prothorax. Larvae were fed small pieces of meat at the burrow's entrance. Once the larva pupated, it was transferred to a dish lined with moist filter paper. The paper was kept moist with 3% hydrogen peroxide solution to prevent fungal infection.

Marlene Palmer (1979) described her success with rearing large numbers of larvae in wooden boxes and glass tubes. She found that almost any container holding soil up to approximately 30 centimeter depth would work to rear larvae, although inert materials such as wood and glass worked best. Glass tubes were found to be an efficient way to keep individual larvae. She used 20-centimeter long (8-inch), 20-millimeter wide (3/4 of an inch) tubes for first and second instar larvae, and 30-cm long (12-in.), 30 mm wide (1 1/4 in.) tubes for third instar larvae. These tubes were filled with soil and one end stopped with water-absorbent foam, or cotton wrapped in cheesecloth. The tubes can be mounted on slanting shelves with the stopped end of the tubes at the downhill end. In a mass-rearing laboratory, these shelves are outfitted so that the tubes can rest in a trough that is partly filled with water to transfer moisture to the tubes (Palmer 1979).

To obtain eggs, Palmer (1979) suggested placing adults in a glass terrarium with moist paper towel as a substrate, since finding and removing eggs from soil is more difficult. Alternatively, oviposition dishes can be made from inverted canning jar rims lined with wire mesh and filled with soil from the site where the adult beetles were collected. Adult females in the terrarium will lay their eggs in this oviposition dish. The soil in the dish should be kept moist.

Newly hatched larvae can be transferred to the glass tubes after they have been removed from their burrows in the field, or from oviposition dishes in the laboratory. A thin grass stem can be placed in the larval burrow, and then carefully excavated until the larva is found. Larvae transferred from the field should be placed in individual containers with some moist soil to prevent desiccation. First instar larvae should be transferred to their laboratory tubes within an hour for best success. Larvae should be cooled down for a few minutes in a refrigerator before they are transferred to their tubes. A hole with a diameter about as wide as or slightly wider than the larva should be prepared in the soil of the glass tube before transferring the larva. Cooled larvae may be gently grasped with forceps on their middle leg and placed abdomen-first into the prepared hole. After the larva is inside the hole, close the entrance by covering it with soil. The larva will reopen the entrance after it has modified the burrow to its satisfaction (Palmer 1979).

Larvae may be fed live insects as well as small pieces of raw hamburger. Place the food into the burrow entrance so you can see it. The larva will usually pull the food down into the burrow (Palmer 1979).

Larvae can be removed from the glass tubes by submerging the entire tube in water for 5 to 10 minutes. After the soil is completely waterlogged and the burrow filled with water, the soil should easily slide out of the tube when the stopper plug is removed. The soil then can be carefully excavated to find the larva. Palmer (1979) had excellent luck with this technique, and never lost a larva (even when accidentally submerging one for 2 days!).

CONSERVATION

Tiger beetles are prime candidates for conservation study by amateur naturalists and research scientists. There are many unanswered questions about tiger beetle ecology. For example: What are the range limits of our species? How far do individuals disperse in the spring and how are new populations established? Can larvae of endangered species be successfully reared and released to establish new populations? Are subspecies evolving into reproductively isolated biological species? Just as the National Audubon Society holds Christmas bird species counts, and the Xerces Society holds Fourth-of-July butterfly counts, tiger beetle counts should become an annual event with naturalists.

Surveying tiger beetle adults is very important to determine the health of local tiger beetle populations. The simplest way to survey is simply to count how many individuals you see in a given area over a given time. This can be easily done for tiger beetles that live in well-defined linear habitats such as on beaches, because their habitat is easy to survey. A beach survey usually is done by two or three people walking abreast and separated from each other by 3 or 4 meters. As you walk along the beach, count how many individuals you see. It is best to do this kind of census at the same time of year (ideally within a few days of the same date) and in the best weather conditions. Although this is a rough estimate, changes in population size can be tracked year after year. It is also important to determine if there is more than one cohort because many species have a 2-or-more-year life cycle. This will affect yearly census data.

Conservation, natural history, and research organizations are in great need of data about where and when different species occur. Data about larvae are particularly important, because the presence of larvae at a site means there is a reproducing population. It is only with this information that endangered species can be identified. It is most important that you record with a photograph or preserved specimen (pinned adults, larvae in alcohol) and include the date and location of species you find. You may wish to join a local entomological society or club. You may find these groups by calling the closest university entomology department or ask people involved with science and stewardship in your local Audubon, Nature Conservancy, or similar office.

GREEN SPECIES

Cicindela sexguttata   40
Cicindela patruela patruela   45
Cicindela scutellaris rugifrons   50

Cicindela sexguttata Fabricius, 1775

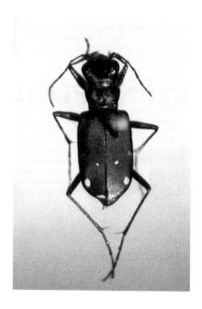

Common Name: Six-spotted tiger beetle.

Adult Identification: 10 to 14 mm in length. Strikingly beautiful metallic blue-green beetle (blue when viewed from an angle) with maculations reduced to 6 white spots along the outer rear edge of the elytra. An unmistakable tiger beetle, and often the first one noticed in the early career of a cicindelaphile. In very few specimens the humeral lunule is represented by a small anterior dot, or in others the remnants of the middle band is present as an extremely thin band and a discal dot. Other specimens have the maculations greatly reduced or absent. The frons is hairless. The labrum has 3 teeth (Fig. 58), the genae are glabrous.

Similar Species: *Cicindela patruela* is superficially similar, but is duller green (because of larger and deeper alveoli) and has a complete middle band. Some individuals of *C. limbalis* and *C. purpurea* are almost as green as *C. sexguttata*, but they have a dark stripe on the lateral margin, a hairy frons, and always show at least some trace of bronze color. *Cicindela scutellaris rugifrons* is green, but the marginal band forms a prominent triangle.

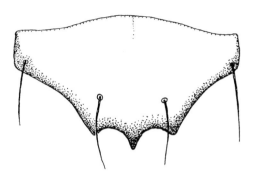

Figure 58. Labrum of adult *Cicindela sexguttata.*

Habitat: A generalist species found in woods on logs and stones, along dirt roads, trails and paths, and in gardens. In most of the U.S., considered to be typical of deciduous woods. In northern areas, such as Vermont, it is not found in deep woods but is common on the edges of the forest, in clearings, gardens, and in recently lumbered areas. Perhaps, more than any other tiger beetle, it is apt to be found during winter hibernation under bark and similar hiding places. In late April these beetles are sometimes found clustered under flat stones or pieces of wood where they can be warmed by the sun without exposing themselves.

Life History: Primarily a spring species, with a 2-year life cycle. Collected as early as March 30 in New England (Larochelle 1986a). Adults are most common in May and June, less so in July. Very rarely found in August because most new adults that emerged from pupae in August remain underground in their pupal cells and overwinter before emerging from their burrows the following spring.

Adult females lay eggs in slightly shaded sand or clay mixed with humus, unlike other tiger beetles that oviposit in sand. Females will lay bright cream-colored eggs under loose leaves (Shelford 1908). Most larvae reach third instar by fall, pupate in midsummer the following year, and may rarely emerge briefly in the southern part of our area at the end of the summer and early fall. Adults hibernate in burrows and emerge sexually mature the following spring.

Third Instar Larva Identification: 20 to 24 mm. The maxillary palpus has three segments. The U-shaped ridge at the caudal end of the frons bears two setae. The first segment of the antenna has 5 or 6 setae, while the second segment has 9 or 10 setae. According to Hamilton (1925), this is the only species in which the dorsal hairs on the head and pronotum are a uniform light chestnut brown color. The head and pronotum are without any trace of metallic color, and the pronotum has no color pattern (as do *Cicindela formosa generosa* larvae). The setal arrangement on the pronotum is shown in Figure 59. The inner hooks have 2 setae (Fig. 60).

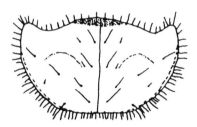

Figure 59. Pronotum of the third instar larva of *Cicindela sexguttata*. (Redrawn with permission from Hamilton 1925.)

Figure 60. Inner hook of *Cicindela sexguttata*. (Redrawn with permission from Hamilton 1925.)

Range: In our area (Figs. 61, 62) found in all New England states, Quebec, New Brunswick, and Nova Scotia. In Canada, found from just west of Lake Winnipeg, through southern Ontario, southern Quebec, east to Nova Scotia (Wallis 1961, Rankin 1996). Found from New England to Florida, west to North Dakota and south to Texas (Pearson et al. 1997). In the midwestern states, an immaculate violet-colored variant is known (Leng 1902).

Notes: There is a form formerly called *Cicindela sexguttata harrisii* Leng, 1902, differing from typical *C. sexguttata* in being duller, darker olive green, with little or no trace of blue tint. It is found from mid- to late summer in forest-edged fields at higher altitudes (above 300 meters elevation) in northern New England. Its true nature has yet to be thoroughly investigated. It could be a local variant (race), or even a separate species, but the unusual color could also be a direct effect of development in cooler temperatures.

Cicindela sexguttata is active at lower temperatures than most other species. Adults often sit in sun flecks and wait for potential prey (Knisley and Schultz 1997).

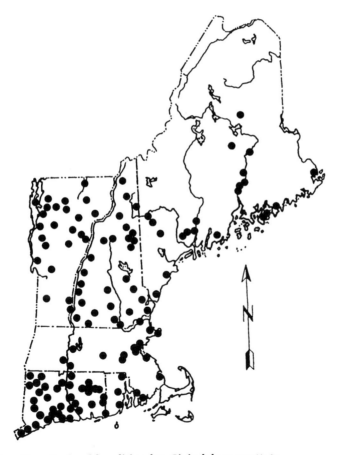

Figure 61. New England localities for *Cicindela sexguttata*.

Figure 62. Eastern Canadian localities for *Cicindela sexguttata*.

Cicindela patruela patruela Dejean, 1825

Common Name: Patterned green tiger beetle.

Adult Identification: 12 to 14.5 mm. A dull-metallic green species (rarely brownish-green), somewhat duller than *C. sexguttata* (because of more and deeper alveoli) with middle band complete. Metallic green on ventral side. A beautiful green beetle with middle band complete, transverse and slightly sinuate, apical lunule reduced to 2 dots, or dots narrowly joined. The humeral lunule is reduced to dots or slightly interrupted, and the marginal band is always absent. The ventral surface is more hairy than in *C. sexguttata*. The area above and lateral to the prosternum (the proepisternum) (Fig. 6) has setae, especially near the coxae. The frons, genae, and clypeus are glabrous. The labrum is white with 3 teeth.

Similar Species: *Cicindela sexguttata* has the white marks reduced to small dots; green individuals of *C. limbalis* and *C. purpurea* have the frons hairy. *Cicindela scutellaris rugifrons* is green, but the middle band is missing and the marginal band is shaped like a prominent triangle.

Habitat: Found on sandy soil where jack pine, oak, blueberry, and sweet fern grow. In Vermont, known from a single specimen collected in

1870 in Burlington. Reported in Ohio along sandy roads, fire lanes, and other coarse-grained sandy areas of eroded sandstone in pine–oak woods (Graves and Brzoska 1991; Knisley and Schultz 1997) and from wooded hill paths (Leng 1902). In Pennsylvania, found in woodland terrain typical for *C. sexguttata* (Howard Boyd, pers. commun.). In Michigan, *C. patruela* is found in the northern forests along sandy little-used roads. Sometimes found in the company of *C. longilabris* (Graves 1963). In Wisconsin, the brown form (*C. patruela huberi* Johnson) is found along grassy lanes in dry jack pine or oak forest (Johnson 1989a). Bousquet and Larochelle (1993) list *C. patruela huberi* as a variety of *C. patruela patruela* and not a true subspecies.

Life History: A spring–fall species with a 2-year life cycle (Shelford 1908). Adults are found from mid-April through the end of June, and again from mid-August into September. In Indiana, adults were much more common in the late summer and early fall than in early summer (Knisley et al. 1990). Willis (1980) reared adults from eggs in a terrarium within 1 year.

Third Instar Larva Identification: 21 mm. Head black with faint green-purple reflections. Diameter of stemma 2 slightly less than stemma 1. The distance between the stemmata is less than the diameter of stemma 2. Antennae (Fig. 63) and base of mandibles red-brown color with the tip of the mandibles black. Pronotum dark orange-brown, lighter around the edge with faint green-purple reflections (Fig. 64). Secondary setae on the pronotum white, other setae yellow-brown color. U-shaped ridge on caudal end of frons with 2 setae. First antennal segment with 7 setae, second segment with 9 to 10 setae. On the abdomen, the sternum of

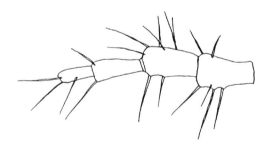

Figure 63.   Antenna of the third instar larva of *Cicindela patruela*. (Redrawn with permission from Willis 1980.)

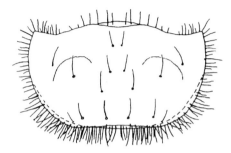

Figure 64. Pronotum of the third instar larva of *Cicindela patruela*. (Redrawn with permission from Willis 1980.)

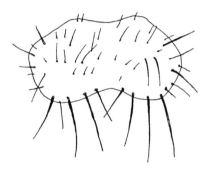

Figure 65. Ninth abdominal sternum of the larva of *Cicindela patruela*. (Redrawn with permission from Willis 1980.)

segment 9 has two groups of 3 long setae and 1 short seta on the caudal margin (Fig. 65). Inner hook with 2 setae on a sloping shoulder, spine is one third to one half the length of the entire hook (Willis 1980).

Range: Records in our area (Figs. 66, 67) include Connecticut, Massachusetts, New Hampshire (two specimens known), Quebec, Rhode Island, and Vermont (Laliberte 1980, Dunn 1981, Bousquet and Larochelle 1993). In Canada, one site is known in southern Ontario (Fig. 67) and another in southwestern Quebec (Wallis 1961). Found in the northeastern and north-central U.S. south to Georgia along the Appalachians (Graves and Brzoska 1991, Pearson et al. 1997).

Notes: A rare treasure in New England that should be photographed but not collected. Only two localities are known in all of Canada, only one specimen is known from Vermont, two from New Hampshire, and a handfull from eastern Massachusetts and Rhode Island (Dunn 1981). A black subspecies (*C. p. consentanea* Dejean, 1825) is known from eastern New York, the New Jersey Pine Barrens (Boyd 1978), and Delaware (Knisley and Schultz 1997), and a brown form (*C. patruela huberi* Johnson) is known from Wisconsin (Lawton 1970, Johnson 1989a).

Tiger beetle students should be alert, as it is entirely possible that populations of this species may still survive in small pockets. In sand country, look along abandoned roads, in old sand pits and other formerly disturbed spots.

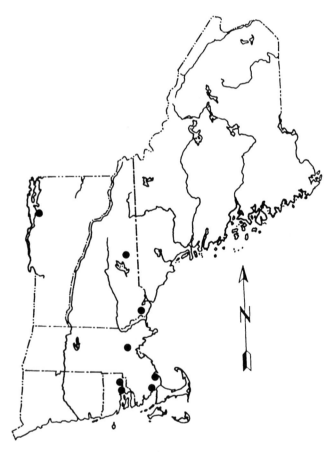

Figure 66. New England localities for *Cicindela patruela*.

Figure 67. Eastern Canadian localities for *Cicindela patruela.*

Cicindela scutellaris rugifrons Dejean, 1825

(Not Illustrated)

Adult Identification: Similar to *Cicindela scutellaris lecontei* except elytral color is deep green to black. Older literature refers to the black color form "modesta" that is now considered to be a synonym of *C. s. rugifrons* (Wilson and Brower 1983, Bousquet and Larochelle 1993).

Similar Species: The only other species that are green are *Cicindela sexguttata, C. patruela*, and the green forms of *C. purpurea* and *C. limbalis. Cicindela scutellaris rugifrons* is the only green tiger beetle with the marginal band represented by a prominent triangle.

Range: Found in Massachusetts, Rhode Island, Connecticut (Fig. 136) and south to Virginia east of the Appalachians (Bousquet and Larochelle 1993).
    (See description of *Cicindela scutellaris lecontei* for complete account of this species.)

DARK SPECIES WITH COMPLETE WHITE MACULATIONS

Cicindela repanda repanda Dejean, 1825

Common Name: Common shore tiger beetle.

Adult Identification: 11 to 13 mm. A common, fully maculate species with a greenish-brown ground color above (partly bronzed under magnification) and metallic blue-green below. The humeral lunule is C-shaped (the posterior end curves inward but not forward); the middle band joins the marginal band. The marginal band is narrowly separated or, more rarely, narrowly joined to the humeral lunule, but is well separated from the apical lunule. The epipleuron (Fig. 6) is pale for most of its length. Elytral microserrations are present. The clypeus is glabrous, while the frons and genae are covered with hairs. The labrum is white with a single dark anterior tooth and 8 setae.

The Canadian subspecies *Cicindela repanda novascotiae* Vaurie 1951 is lighter and more bronze in color and has reduced or partially obliterated maculations compared with *C. repanda repanda* (Vaurie 1951).

Similar Species: The well-developed C-shaped humeral lunule and pale epiplura of *C. repanda repanda* separates this species from other brownish ones. *C. repanda novascotiae* could easily be confused with *C. duodecimguttata*, where the marginal band is interrupted or absent and the maculations are reduced to isolated round dots or are very narrowly connected. *Cicindela duodecimguttata* can be separated from *C. repanda* by the dark epiplura, the wide pronotum, and (from *C. repanda repanda*) the reduced maculations. *Cicindela repanda repanda* has a complete marginal band (which may be very narrow). In *C. hirticollis hirticollis,* the genae are glabrous (hairy in *C. repanda*) and the posterior arm of the humeral lunule turns anteriorly, so the lunule appears like a capital G. In *C. ancocisconensis,* the posterior end of the humeral lunule is oblique and thickened so that it appears like the decimal 7, and the labrum has 3 teeth (rather than 1, as in *C. repanda*).

Habitat: A generalist species found in a variety of habitats, but most abundant near water. Adults are very abundant on sandbars and mud flats along rivers, though they also occur far from water in sandpits and along sandy roads. A record from 1490 m elevation (4900 ft.) near the top of Mt. Franklin, NH, represents an individual that was probably swept there by rising air currents (Dunn 1981). Larvae prefer heavy or sandy moist soil along riverbanks, mud flats, and sandy bars. *Cicindela duodecimguttata* is found in similar habitats but is more often found farther away from the water's edge (Wallis 1961).

Life History: There are collection records for this species from March through October in many states. In our area this spring–fall species is common through the summer season, except in July when numbers are much lower. The life cycle is usually complete in 2 years. In New Hampshire, mating has been observed in late May, early June, and late July (Dunn 1981).

Females oviposit in early summer, and the larvae enter into hibernation the following fall as third instars. The eggs are bright yellow and are laid in sloping sand in May and June. The larvae pupate and emerge as sexually immature adults the following summer, then pass through another winter in a hibernaculum burrow to emerge sexually mature the following May (Hamilton 1925).

Third Instar Larva Identification: 16 to 18 mm. The head and pronotum are dark copper bronze with light green reflections. Median hooks with one large seta (or if two, one is very small). The first antennal segment has 9 to 10 setae, the second segment has 7 or 8 setae. There are few secondary setae on the pronotum (Fig. 68). Inner hook has 2 setae on the shoulder and the central spine projects only one sixth or

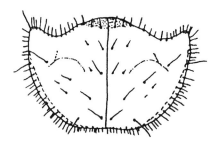

Figure 68. Pronotum of the third instar larva of *Cicindela repanda*. (Redrawn with permission from Hamilton 1925.)

Figure 69. Inner hook of *Cicindela repanda*. (Redrawn with permission from Hamilton 1925.)

less the entire length of the hook (Fig. 69). U-shaped ridge on the frons has 2 distinct setae (Hamilton 1925).

Larvae of *C. repanda* have been found burrowing in more diverse habitats than any other species (including soil with humus, mud, wet sandy soil, and moist clay). However, they are most commonly found burrowing on the edges of moist vegetated soil bordering ponds, streams and rivers. Burrows are usually 15 cm deep (6 in.) and slope obliquely into the soil on level ground and are perpendicular to a sloping soil surface (Phil Nothnagle, pers. commun.). Larvae will leave their burrows if the soil becomes dry (Hamilton 1925).

Range: Widely distributed. Found throughout our area (Figs. 70, 71) (Bousquet and Larochelle 1993). Across Canada from southern British Columbia to Newfoundland. Occurs North to Fort McMurray in Alberta, Cape Henrietta Maria on Hudson's Bay, and Hopedale in Labrador (Wallis 1961). Found throughout the U.S., except in parts of the Great Basin, California and Nevada (Graves and Brzoska 1991, Pearson et al. 1997).

Notes: *Cicindela repanda* has a characteristic short erratic flight behavior (Graves and Brzoska 1991).

A difficult species to rear; larvae will abandon burrows if conditions are not perfect. Wandering larvae will be eaten by others if reared together in the same terrarium. One of the most common species found in our area.

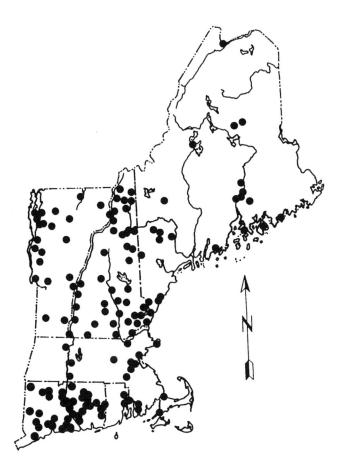

Figure 70. New England localities for *Cicindela repanda.*

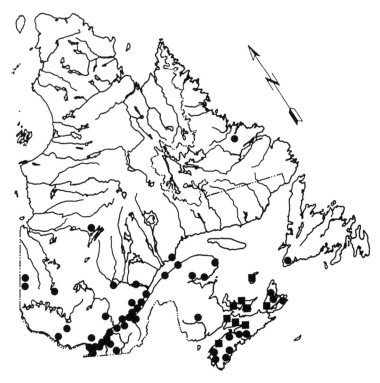

Figure 71. Eastern Canadian localities for *Cicindela repanda*: (■) *C. repanda novascotiae*; (●) *C. repanda repanda.*

Cicindela formosa generosa Dejean, 1831

Common Name: Big sand tiger beetle.

Adult Identification: 16 to 18 mm. Our largest *Cicindela* species. In eastern North America, *C. formosa generosa* is easily distinguished from our other species by its large size and overall robust appearance, and the complete maculation pattern where the lunules and marginal band are all connected along the lateral margin of the elytra. A reddish-bronzed to brown tiger beetle above and metallic green to cupreous below. The posterior end of the humeral lunule can be oblique (resembling a thick version of *C. tranquebarica*) or C-shaped. The frons and genae are covered with setae, while the clypeus is glabrous. The labrum is 3-toothed with outer teeth slightly splayed laterally. Elytral microserrations very fine. First antennal segment with 12 or more setae.

Similar Species: *Cicindela formosa* is a variable species found across North America. In western Canada for example, the subspecies *C. formosa gibsoni*, can be so completely maculate that the elytra appear almost totally white (Wallis 1961; Plate 1). *C. tranquebarica* is another bronzed species but is less reddish, and has narrower markings which are not connected along the margin.

Habitat: *Cicindela formosa* is a dry-habitat species that is found on yellow-to-white shifting sand with sparse vegetation. Found in the short grass and weeds near the edges of sand dunes (Leng 1902); in old, seldom-used road cuts, sand pits and seashores; and in pine barrens. Inland often found with *C. scutellaris lecontei.*

Life History: A spring–fall species with a 2-year life cycle. Found from late April through July, rarely in August, and again in late August through September. Adults from two different broods may overlap in midsummer. Mating pairs seen in New Hampshire July 5 and 7 (Dunn 1981).

Female beetles mate and lay eggs in June to early July in slightly shifting fresh sand. The larvae reach third instar in September, close their burrows by October, and overwinter. In the spring, the larvae open their burrows, then pupate in a chamber on the side of the burrow tunnel about 10 cm (4 in.) below the soil surface. Some adults may emerge in the late summer, but many remain in their pupal chambers to overwinter. Sexually mature adults reemerge in late April through May (Hamilton 1925).

Third Instar Larva Identification: 22 to 24 mm. The head and pronotum are chestnut-brown color (similar to *Cicindela sexguttata*), but there is a color pattern formed by lighter areas (Fig. 72). Setae on the top of the head and pronotum are clear to white, while the rest of the body setae are brown. The U-shaped ridge on the back of the frons has 3 setae. The 1st antennal segment has 6 or 7 setae while the 2nd segment has 9 or 10. On the abdomen, the sternum of segment 9 has two groups of 4 setae on the caudal margin. The median hook has 3 setae while the inner hook has 4 setae (Fig. 73) (Hamilton 1925).

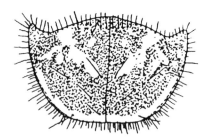

Figure 72. Pronotum of the third instar larva of *Cicindela formosa generosa*. (Redrawn with permission from Hamilton 1925.)

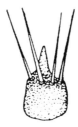

Figure 73. Inner hook of *Cicindela formosa generosa*. (Redrawn with permission from Hamilton 1925.)

Burrows of *C. formosa generosa* are dug in slightly shifting sand, along the edge of sand dunes and forest habitat. The shape of the burrow is distinctive (Shelford 1908: Plate 25, Fig. 24). The burrow is vertical for most of its length (between 30 to 50 cm [12 to 20 in.] deep), but 2 cm or so below the soil surface, the burrow turns to a horizontal position and opens into a small pit. Hamilton (1925) says this pit functions as a trap for insect prey, and also to catch and prevent sand from filling up the burrow. The larva cements the sand at the mouth of the burrow with saliva.

Range: In our area (Figs. 74, 75), known from all the New England states and Quebec. In Canada, found in southern Ontario (Wallis 1961). *Cicindela formosa* is widely distributed from New England and the middle Atlantic states, south to Texas and north to Alberta (Pearson et al. 1997).

Notes: A wary species that flies when disturbed and makes a distinct buzzing sound during flight (Graves and Brzoska 1991).

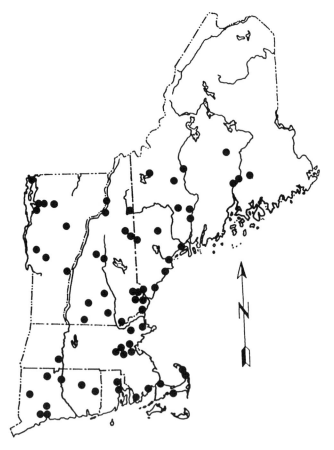

Figure 74. New England localities for *Cicindela formosa generosa.*

Figure 75. Eastern Canadian localities for *Cicindela formosa generosa*.

Cicindela ancocisconensis T. W. Harris, 1852

Common Name: White mountain tiger beetle.

Adult Identification: 14 to 16 mm. A brownish-bronzed species with greenish-blue color on the top of the head, thorax, and abdomen. Elytra with complete white maculations including a well-developed marginal band. The head, thorax, and abdomen are metallic blue-green to violet below. The frons and genae are hairy while the clypeus is glabrous. The posterior end of the humeral lunule is abbreviated and slightly oblique (like the digit 7 with a thick base). The labrum has 3 teeth with the outer 2 often slightly splayed laterally. First antennal segment with 9 or more setae.

Similar Species: *Cicindela repanda* is superficially similar, but has a C-shaped humeral lunule and one tooth on the labrum. *Cicindela tranquebarica* has the end of the humeral lunule longer, thinner, and more oblique than in either *C. repanda* or *C. ancocisconensis*. Also, the marginal band is scarcely developed in *C. tranquebarica*.

Habitat: Found in shaded gravel, sandbanks, and sandbars on mountain brooks and medium-to-small rivers with large boulders and shade. Probably widespread through the mountains.

Life History: A spring–fall species with a 2- or 3-year life cycle (Knisley and Schultz 1997), found from May and June, rarer in midsummer, then common again in mid-July into early September (Dunn 1981).

Third Instar Larva Identification: The larvae are undescribed. However, larvae were collected along with adults in West Virginia by Bob Acciavatti (Carnegie Museum of Natural History) and loaned to us. Some of the larvae were reared to adult *C. ancocisconensis*. The following description is from the above specimens:

18 to 20 mm. The head and pronotum are dark brown to black with faint coppery reflections. Setae on head and pronotum (Fig. 76) are white, while the body setae are brown. There are 2 setae on the U-shaped ridge on the back of the frons. The 1st antennal segment has 4 setae, while the 2nd segment has 7. Stemma 1 is slightly larger than stemma 2, and the distance between them is slightly less than the diameter of stemma 2. On the abdomen, the sternum of segment 9 has 2 groups of 3 large and 1 small seta along the caudal margin. The median hook has 4 setae. The inner hook (Fig. 77) as 2 to 3 setae on the shoulder and 1 seta on the base. The setal plan of the 3rd abdominal segment is shown in Figure 78.

Larval burrows are found in sandy-loam soil away from the water's edge (Wilson 1979).

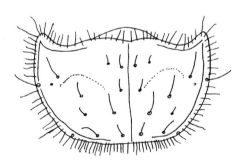

Figure 76. Pronotum of the third instar larva of *Cicindela ancocisconensis*.

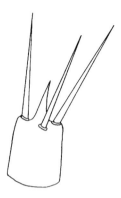

Figure 77. Inner hook of the larva of *Cicindela ancocisconensis.*

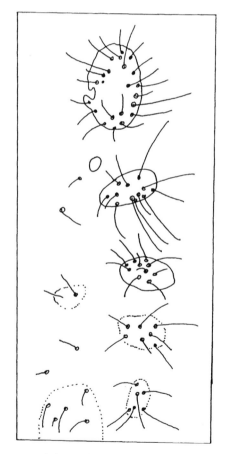

Figure 78. Setal pattern of the 3rd abdominal segment of the third instar larva of *Cicindela ancocisconensis.*

Range: In our area (Figs. 79, 80), found in Massachusetts, Maine, New Hampshire, New York, Quebec, and Vermont (Wilson and Brower 1983, Nelson and Labonte 1989, Bousquet and Larochelle 1993, Pearson et al. 1997). Recorded from Gaspé Provincial Park, Quebec (Wallis 1961). South to Georgia along the Appalachian mountains, and west to Indiana and Illinois (Bousquet and Larochelle 1993).

Notes: Probably a widespread species. Because little is known about this beautiful beetle, it deserves more study. It has not been found in recent years in Ohio (Graves and Brzoska 1991) and may be endangered in parts of its former range.

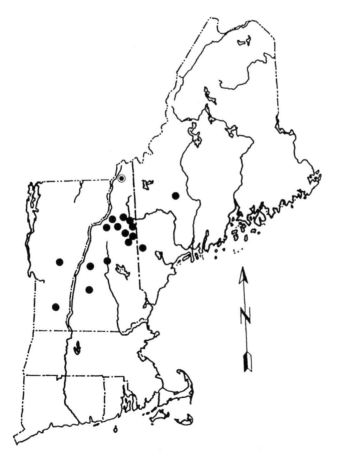

Figure 79. New England localities for *Cicindela ancocisconensis.* ◉ Indicates population with reduced middle band in Pittsburg, NH.

Figure 80. Eastern Canadian localities for *Cicindela ancocisconensis*.

In northern New Hampshire (Pittsburg, Coos County), there is a population with an interrupted "arm" of the middle band (Fig. 81), thus creating a marginal line and a discal dot (Fig. 79) (Dunn 1979, 1981).

Figure 81. Elytral markings showing interrupted middle band of *Cicindela ancocisconensis* from Pittsburg, NH. (Redrawn with permission from Dunn 1979.)

Philip J. Darlington, Jr. (pers. commun.) noted that this species often occurs together with *Cicindela repanda*. When disturbed in the field, *C. ancocisconensis* flies higher and farther in a straight line than *Cicindela repanda*. Therefore, it may escape notice when both species are present.

T. D. Harris chose the name "ancocisco" because John Smith used this name to refer to the "twinkling mountain" or White Mountains of New Hampshire (the type locality). Further research, however, has shown that Smith probably had a miscommunication with the local Native American interpreter, because according to Eckstorm (1941) the word "ancocisco" means "muddy bottom" (Wilson 1979).

Cicindela tranquebarica tranquebarica **Herbst, 1806**

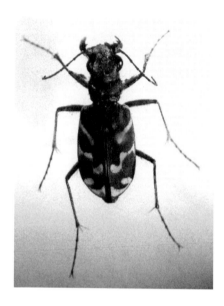

Common Name: Oblique-lined tiger beetle.

Adult Identification: 13 to 16 mm. A <u>dark</u> bronzed brown-black species with a hairy frons, easily identified by the <u>long oblique end of the humeral lunule</u>. Marginal band is very thin and short. Head and thorax coppery to greenish-bronze below, while the abdomen underneath is metallic blue-green. Elytral microserrations are present. The 1st antennal segment has 8 or more setae. The clypeus and genae are glabrous, while the frons is hairy. The labrum has 3 teeth.

Similar Species: The only two other species with oblique humeral lunules are *C. ancocisconensis* and some individuals of *C. formosa generosa*. Both of these have well-developed marginal bands. *C. f. generosa* is a large robust tiger beetle with the marginal band connecting the humeral and apical lunules. The marginal band is present in *C. ancocisconensis* but does not connect the humeral and apical lunules, and the posterior end of the humeral lunule is short and stubby. *C. ancocisconensis* has setae on the genae while in *C. tranquebarica* the genae are glabrous.

Habitat: A generalist species with habitat quite varied, associated at times with most of the other species in our area: on clay or sand, including sand and gravel pits, gravel roads, eroded areas, beaches, and dune sand. Found in upland woods on dark-colored gravel roads.

Life History: A spring–fall species with a 2-year life cycle. A harbinger of spring and one of the last species to be seen before winter. Collected on February 27 in Ohio (Graves and Brzoska 1991) and on March 7 in New Jersey (Boyd 1978) and March 29 in New Hampshire (Dunn 1981). Adults most common in late April and May, and again in September and October. In New Hampshire, mating observed on May 31, June 6, and July 20 (Dunn 1981).

Female adults lay eggs in sandy, vegetated soil with some humus starting in late April. The larvae feed, grow, and reach third instar by fall hibernation time. The larvae emerge and feed again the following spring and pupate in early to midsummer. Sexually immature adults emerge in August and hibernate in the fall and winter. The following spring they emerge sexually mature.

Third Instar Larva Identification: 21 to 24 mm. The maxillary palpus has 3 segments. The U-shaped ridge on the back of the frons has 2 distinct setae. The 1st segment of the antenna has 7 or 8 setae, while the 2nd segment has 9 to 10 setae. There are 2 setae on the shoulder of the inner hook (Fig. 82). The median hook has 3 setae. The 9th abdominal sternum has two groups of 4 setae each on the caudal margin. The distance between stemma 1 and 2 is less than the diameter of stemma 2. The head and pronotum are bright coppery-bronze to purple-bronze color with strong bluish-green reflections. Setae on head and pronotum (Fig. 83) are white, while body setae are brown. The burrow ranges from 22 to 50 cm deep (9 to 20 in.) and is straight (Shelford 1908).

Figure 82. Inner hook of *Cicindela tranquebarica*. (Redrawn with permission from Hamilton 1925.)

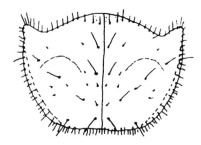

Figure 83. Pronotum of the larva of *Cicindela tranquebarica*. (Redrawn with permission from Hamilton 1925.)

Adult females oviposit in soil where the water table is relatively close to the surface, unlike *C. formosa generosa* or *C. scutellaris*, which oviposit in well-drained soil (Knisley and Schultz 1997).

Range: Throughout our area (Figs. 84, 85) (Bousquet and Larochelle 1993). Across Canada from British Columbia, just south of Great Slave Lake in the Northwest Territories, eastward to southern Ontario, southern Quebec, Nova Scotia, Sable Island, and western Newfoundland (Wallis 1961, Rankin 1996). In Canada, *C. tranquebarica* is the second most common tiger beetle after *C. repanda* (Wallis 1961: 57). Found in all states of the lower 48 except perhaps Florida (Pearson et al. 1997). One of the most widely distributed tiger beetles in North America.

Notes: *Cicindela tranquebarica* is capable of long sustained weaving flight; a difficult beetle to photograph or catch.

The larvae of *C. tranquebarica* are often parasitized by the bee fly *Anthrax* (Knisley and Schultz 1997).

The species name (*tranquebarica*) comes from a forgotten Danish colony in southern India. Huber (1986) suggested that Herbst may have accidentally switched labels between a North American species (*tranquebarica*) from Baltimore, MD, and an Indian specimen (*baltimorensis*) from Tranquebar, because Herbst's *baltimorensis* is a synonym of the Oriental *Cicindela minuta* Olivier. (This illustrates the fact that scientific names of species are never changed on grounds of inappropriateness, but only when the rules of nomenclature are found to have been violated. This is a good principle; otherwise, there would be endless arguments over appropriateness.) This species also used to be called *Cicindela vulgaris*, which means "common Cicindela," a suitable name, which, alas, is illegal under the rules.

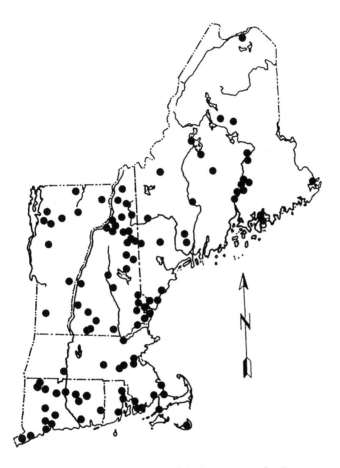

Figure 84. New England localities for *Cicindela tranquebarica.*

Figure 85. Eastern Canadian localities for *Cicindela tranquebarica*.

Cicindela puritana G. H. Horn, 1871

Common Name: Puritan tiger beetle.

Adult Identification: 12 to 14 mm. A greenish-bronzed species, more parallel-sided, elongate, and slender than *C. repanda*. The elytron has a distinct lateral subapical angle (Fig. 86). This is obtuse in the male, with the margin behind it slightly sinuate. In the female there is a rectangular angle followed by a deep emargination. The pattern of white maculations is extensive, similar to *C. marginata* or western *C. hirticollis*, but more obvious and with all lunules connected by a wide marginal band. The humeral lunule is strongly G-shaped, with the inner anterior arm enlarged and recurved. The middle band is recurved and ragged. The anterior arm

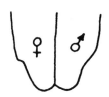

Figure 86. Elytral shape of female and male *Cicindela puritana*. (Redrawn with permission from Leng 1902.)

of the apical lunule is directed almost anteriorly (oblique in other species). The clypeus, frons, and genae, along with the thorax and femurs, are covered with white decumbent setae. The prosternum is glabrous. The 1st antennal segment bears 2 setae.

Similar Species: No other species in our area have the angulation on the elytron. *Cicindela hirticollis* has a similar G-shaped humeral lunule, but the details of the shape are different (see Plate 2). *Cicindela hirticollis* also has the middle band much less recurved and scarcely ragged. *Cicindela marginata* has a ragged recurved middle band, but is a more robust beetle with apical ends of the elytra with a strong declivity (in females), and males have a mandibular tooth, not at all like *C. puritana.*

Habitat: New England populations prefer wide sand deposits along big rivers or narrow beaches along rivers with clay banks (U.S. Fish and Wildlife Service 1993a). Only one specimen is known from Vermont, taken at Hartland along the Connecticut river. Years ago known from the New Hampshire side of the Connecticut river from Claremont, NH to near the river's mouth in Connecticut (Leng 1902). Now limited to a few small populations along the river.

Museum records from Massachusetts include the towns of Chicopee, Hadley, Longmeadow, and Springfield. In Connecticut, records include the towns of Hartford, Windsor, and Portland.

Life History: Probably a summer species with a 2-year life cycle. Museum of Comparative Zoology, Harvard University, records for New England show collection dates from August 4 to September 30 (Larochelle 1986a). Recent work by Philip Nothnagle (U.S. Fish and Wildlife Service 1993a) records adults emerging in July and disappearing by the end of August. Leng (1902) cites collection dates from June 20 to August 13 that undoubtedly include more southerly collection localities.

Third Instar Larva Indentification: The larva of *C. puritana* has been described (as follows) by C. Barry Knisley in an undated report to the Massachusetts Natural Heritage Program:

11 to 18 mm. The mandibles and labrum are brown to yellow-brown. The U-shaped ridge on the back of the frons has 2 setae. The 1st segment of the antenna has 7 to 9 setae, while the 2nd segment has 10 to 13 setae (usually 11 to 12). The distance between stemma 1 and stemma 2 is approximately twice the diameter of stemma 1. The head and labrum are

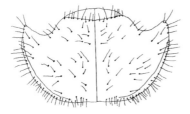

Figure 87. Pronotum from third instar *Cicindela puritana* larva. (Redrawn with permission from Knisley undated.)

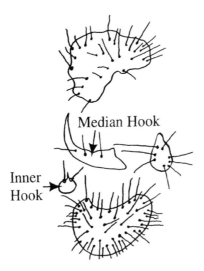

Median Hook

Inner
Hook

Figure 88. Median and inner hooks of third instar *Cicindela puritana* larva. (Redrawn with permission from Knisley undated.)

shiny metallic black. Setae on head and pronotum (Fig. 87) are transparent and glassy and twice as numerous as on *C. repanda*. There are 2 setae on the shoulder of the inner hook (Fig. 88) and 2 to 3 additional setae lower down on the shoulder. The spine of the inner hook is minute, about one sixth the length of the hook. The median hook is long and has 3 setae (Fig. 88). The 9th abdominal sternum has two groups of 3 setae on the caudal margin (Fig. 89). The body setae are yellow to golden brown. The setal arrangement of a lateral view of the 3rd abdominal segment is shown in Fig. 90.

The New England population of larvae burrow in sandy beaches with vegetation and occasionally near the water's edge or in the face of the riverbank (U.S. Fish and Wildlife Service 1993a). *Cicindela puritana* larvae usually burrow between 30 and 70 cm (12 to 28 in.) deep, compared

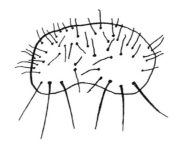

Figure 89. Caudal sternum of abdominal segment 9 of third instar *Cicindela puritana* larva. (Redrawn with permission from Knisley undated.)

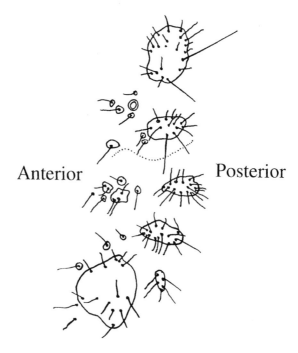

Anterior   Posterior

Figure 90. Lateral view of abdominal segment three of third instar *Cicindela puritana* larva showing setal plan. (Redrawn with permission from Knisley undated.)

with the 15 cm (6-in.) average depth of *C. repanda* larvae. Many larvae overwinter as 3rd instars (Philip Nothnagle, pers. commun.).

Range: Known only from Connecticut, Delaware, Massachusetts, Maryland, New Hampshire, New Jersey, New York, Virginia, and Vermont

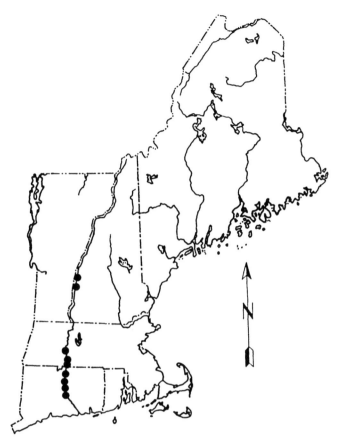

Figure 91. New England localities for *Cicindela puritana*.

(Bousquet and Larochelle 1993; Pearson et al. 1997; Howard Boyd, pers. commun.). In New England, presently known only from a few localities along the Connecticut River (Fig. 91) (U.S. Fish and Wildlife Service 1993a).

Cicindela puritana is considered a sister species of *C. cuprascens* and *C. macra* — these latter two species being found in the midwestern and southwestern states of the U.S. (Pearson et al. 1997).

Notes: *Cicindela puritana* was listed as a threatened species by the U.S. Fish and Wildlife Service in 1990. In addition, *C. puritana* is listed endangered in the states of Massachusetts and Connecticut.

The disappearance of many of the historical sites of our New England populations of *C. puritana* is associated with the construction of flood control and hydroelectric dams along the Connecticut River after World

War II. The last collection records from New Hampshire are from the late 1920s. Our remaining New England populations declined between 1988 and 1993, due to increased recreational usage of beaches (U.S. Fish and Wildlife Service 1993a), but have since shown some increases (Philip Nothnagle, pers. commun.).

A gregarious species that can be found in high densities in the right habitat. Adults are active during the day and at night. The tiphiid parasitic wasp *Methocha* has been found in the burrows of second and third instar *C. puritana* larvae (Knisley and Schultz 1997).

Cicindela hirticollis hirticollis Say, 1817

Common Name: Hairy-necked tiger beetle.

Adult Identification: 13 to 14 mm. A brownish-bronzed species with complete maculations. Similar in size to *C. repanda* but broader, especially the female. The posterior end of the humeral lunule curved anteriorly at the end, suggesting the capital letter "G." The ventral side of this beetle species is metallic blue-green on the abdomen to coppery red-purple on the thorax and head, and covered with long white hairs. Adults have hairy legs and a strikingly hairy thorax. Elytral microserrations are present. The frons has many white setae, the clypeus is bare and has many fine wrinkles. The genae are glabrous, but may have extremely fine longitudinal setae visible in high magnification. The first antennal segment has the usual 3 sensory setae and at most 2 extra setae, and is otherwise glabrous. The labrum is striking: ivory white with a thin black anterior margin and a single dark tooth. Legs are long, shiny metallic green, and covered with stiff white hairs.

At a distance *C. hirticollis hirticollis* may be recognized by its long legs, which it uses to stand higher off the beach surface than the other species with which it is often associated.

Similar Species: At first glance in the field, these beetles appear like a dark version of *C. repanda*; however, the humeral lunule of *C. hirticollis* suggests a capital letter "G," while in *C. repanda*, this lunule is C-shaped. *C. repanda* is not nearly as hairy as *C. hirticollis*. Genae glabrous in *C. hirticollis*, hairy in *C. repanda*.

The subspecies *Cicindela hirticollis hirticollis* is distinguished from *C. h. rhodensis* by having complete maculations and a smaller average size (Graves et al. 1988). Intergrades may be found.

Habitat: *Cicindela hirticollis hirticollis* is a wet-habitat species found on wet sand beaches next to large bodies of fresh or salt water where there are sand dunes nearby. In New England, known only from Lake Champlain, VT, where it is found on the sandy beaches formed at the mouths and shores along large rivers emptying into the lake. Larvae burrow in moist sand in protected salt- and freshwater bays where there is little wave action.

Life History: A spring–fall species with a 2-year life cycle. May have a 1-year life cycle farther south. Found from mid-May through September, but most abundant in June, and August. Female beetles lay eggs in moist clean sand in late June or early July. The larvae reach third instar before closing the burrow in the fall and overwintering. The larvae reopen the burrow in May, feed, and pupate in June and July. Sexually immature adults may emerge in late July and into August, then overwinter again to reemerge sexually mature the following spring.

In late July, newly emerged beetles and year-old adults are present together because adjacent-year broods overlap. Between late July and mid-August more of the adults present are newly emerged, while the year old brood dies off (Shelford 1908).

Third Instar Larva Identification: 17 to 19 mm. The head and pronotum have a bright copper color with metallic green reflections. Setae on the top of the head and the pronotum are white and flattened (on the pronotum) while the other body setae are brown. The most distinguishing character for identifying this species is the large number of white flattened or sack-like setae on the pronotum (Fig 92). The U-shaped ridge at the back of the frons has 3 setae. The 1st antennal segment has 7 to 9 setae, the 2nd segment has 10 to 12. On the abdomen, the median hook has 2 setae, the inner hook has 2 setae. The projecting spine on the inner hook

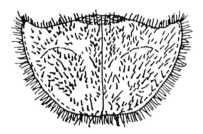

Figure 92. Pronotum of the third instar larva of *Cicindela hirticollis*. (Redrawn with permission from Hamilton 1925.)

Figure 93. Inner hook of *Cicindela hirticollis*. (Redrawn with permission from Hamilton 1925.)

is about one fourth the length of the entire hook (Fig. 93). On the 9th abdominal segment, the sternum has 2 groups of 3 setae (Hamilton 1925).

The burrows are dug in moist, clean sparsely vegetated, slightly shifting sand on beaches. The burrows go straight down and are rarely more than 5 in. deep (15 to 20 cm) (Shelford 1908). Larvae will abandon their burrows if they become too dry or wet. Barry Knisley has observed *C. hirticollis* larvae crawling away from the water to establish new larval burrows further up the beach (Knisley and Schultz 1997).

Range: Along Lake Champlain, VT, *C. hirticollis* is fully maculate and is the nominate subspecies. *Cicindela hirticollis* was once found throughout our area (Figs. 94, 95) and west to the great lakes (Graves et al. 1988, Bousquet and Larochelle 1993), but has disappeared from many of its former localities. The species as a whole has a disjunct range distribution because of its narrow habitat requirements. It has been found from Carter Basin, Labrador, south to Rhode Island, including the Magdalen Islands, Miquelon, Sable Island, and Gardiners Island, NY. In Long Island, NY, *C.*

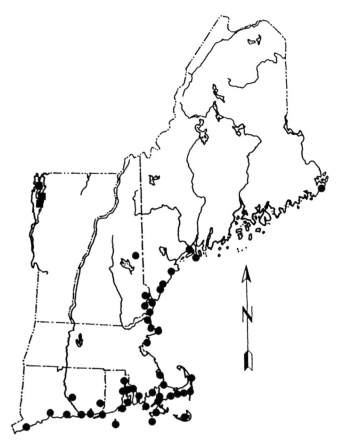

Figure 94. New England localities for *Cicindela hirticollis*: (■) *hirticollis*; (●) *rhodensis*.

hirticollis hirticollis intergrades into the more northerly coastal subspecies *C. hirticollis rhodensis* (Graves et al. 1988).

  Cicindela hirticollis in the broad sense is widely distributed across North America with many subspecies. It is found from the shores of the Pacific to the Atlantic. In Canada, *C. hirticollis* is found from south-central British Columbia, north to Athabasca, and across the continent to Labrador, Newfoundland, and Sable Island off Nova Scotia (Wallis 1961, Lindroth 1954a, Lindroth 1954b, Graves et al. 1988, Rankin 1996). Found south to Mexico and Florida (Graves 1988, Pearson et al. 1997).

  The strength of the elytral maculation pattern varies geographically where darker, less maculated populations are found in the northeast

Figure 95. Eastern Canadian localities for *Cicindela hirticollis,* subspecies not specified.

(Graves et al. 1988). Some populations may be almost completely without maculations (Graves 1963).

Notes: A dwindling species that has disappeared from many localities where it was formerly found. In Vermont there are only five historical sites along Lake Champlain where *Cicindela hirticollis hirticollis* has been collected. The tiger beetle habitat of at least three of these sites has been destroyed through building and human recreation.

In southern California, the three remaining populations of the subspecies *C. hirticollis gravida* are in imminent danger of extinction (Nagano 1980; Graves et al. 1988).

This is a difficult species to rear, because larvae abandon burrows if the moisture level is not perfect (Shelford 1908).

Larvae are often attacked by *Anthrax* (Diptera: Bombyliidae) or *Methocha* (Hymenoptera: Tiphiidae) (Knisley and Schultz 1997).

Cicindela marginata Fabricius, 1775

Common Name: Salt marsh tiger beetle.

Adult Identification: 11 to 14 mm. An attractive brown-bronzed to green-bronzed beetle above, metallic green below. Overall appearance sleek. Marginal band complete along outer edge of elytra joining the lunules. The foot of the middle band broad and with the appearance of being shot through with many small holes. Body underneath (including the prosternum) covered with fine white decumbent hair. Genae, clypeus, and frons covered with fine white decumbent hair. Long slender legs with tan to red trochanters. Apex of elytra of female is deflexed (Fig. 96), while in males there is an apical tooth on the elytra. Males show a prominent tooth on the right mandible (Fig. 97). Elytral microserrations are present but very fine. The 1st antennal segment has 2 setae.

Similar Species: The only other brown tiger beetle found in or near the same habitat is *C. hirticollis*, but *C. marginata* has the broad "Swiss-cheese" foot of the middle band, and the beetles are found most often on dark mud-flats, where *C. hirticollis* is found on the moist sandy beach. Only *C. marginata* has red trochanters and a deflexed elytral apex (females) or prominent mandibular tooth (male).

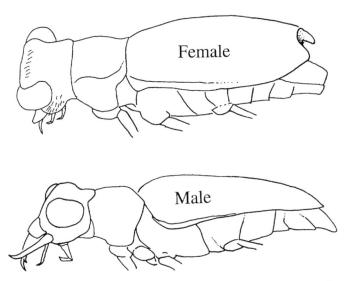

Figure 96. Profile of female and male *Cicindela marginata*. Note the deflexed elytral apex of the female.

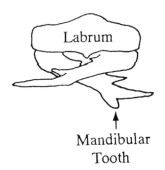

Figure 97. Labrum and mandibular tooth of adult male *Cicindela marginata*. (Redrawn with permission from Willis 1968.)

Habitat: A mud-flat species found in the salt marsh on the bay side of barrier beaches.

Life History: A summer species. In our area adults found from late July to mid-September (Larochelle 1986a). Further south it is a spring to early summer species, found from late April through July (Leng 1902). Mating observed in New Hampshire July 31 (Dunn 1981). The details of *C. marginata*'s life history are unknown.

Burrows can be found near the shore in sandy vegetated soil and are approximately 25 cm (9 to 10 in.) deep.

Third Instar Larva Identification: 19 to 22 mm. The head and pronotum are dark purple-bronze color, with distinct blue reflections. Setae on the top of the head and pronotum (Fig. 98) are transparent, while the other body setae are brown. The U-shaped ridge on the back of the frons has 2 setae. The first segment of the antenna has 9 or 10 setae, as does the 2nd segment. On the abdomen, the 9th sternum has two groups of 3 setae on the caudal margin. The median hook has 3 setae, while the inner hook has 9 or 10 setae (Fig. 99).

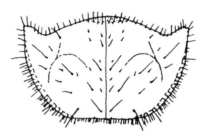

Figure 98.   Pronotum of the larva of *Cicindela marginata*. (Redrawn with permission from Hamilton 1925.)

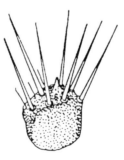

Figure 99.   Inner hook of *Cicindela marginata*. (Redrawn with permission from Hamilton 1925.)

Range: In our area (Fig. 100) found in Massachusetts, Maine, New Hampshire, and Rhode Island. Not known from Canada. Found south along the Atlantic coast to Florida (Bousquet and Larochelle 1993, Pearson et al. 1997).

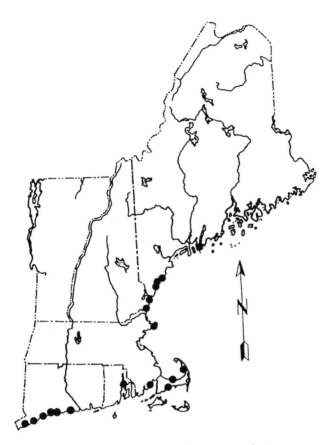

Figure 100. New England localities of *Cicindela marginata*.

Notes: A dwindling species, as is *C. hirticollis*, which is often found with or near *C. marginata*. In New Hampshire, this once-common beetle had been reduced to one population in Rye (Rockingham County). In 1977, this population was in the target area of heavy insecticide spraying for biting flies; no beetles have been seen since.

The adults of *C. marginata* may be collected by flashlight at night (Dunn 1981).

Cicindela limbata labradorensis Johnson, 1990

Common Name: Goose Bay tiger beetle.

Adult Identification: 10 to 12 mm. Head bright bronze with green reflections. Frons with many punctures and covered with erect hairs, especially near the eyes. First antennal segment (scape) with 2 to 4 sensory hairs. Labrum white with a small central tooth. Elytra with microserrations at apical end. <u>Maculations very broad. The humeral lunule is long and wide as is the middle band and the apical lunule. The humeral lunule either merges with, or almost joins, the middle band</u>. The marginal band most often complete, but may be briefly interrupted near the humeral lunule. Underneath the beetle is bright green with a bronze tinge and covered with white decumbent hairs. The genae have setae (Johnson 1989b).

Similar Species: Because of its limited range, *Cicindela limbata labra-dorensis* is unlikely to be mistaken as any other tiger beetle. It superficially resembles *C. formosa generosa* in having bold maculations, but in *C. l. labradorensis* the humeral lunule is so broad that it touches or almost touches the middle band.

Habitat: Extremely limited. Known only from sandy, human disturbed areas such as roads and drainage ditches immediately around the Goose Bay airport, in Labrador. However, tiger beetles have not been extensively collected in Labrador, as sand areas are often inaccessible. In western species of *Cicindela limbata*, beetles prefer sand dunes away from water (Johnson 1989b).

Life History: Probably a 3-year or more life cycle, because *C. limbata* in Manitoba takes 3 years to reach adulthood (Hamilton 1925).

Third Instar Larval Identification: Cicindela *limbata labradorensis* larvae are undescribed, but *C. limbata* larvae from Colorado were described by Hamilton (1925: 56) and are summarized as follows: 15 to 17 mm. Head and pronotum (Fig. 101) coppery bronze with blue-green reflections. Setae on top of head and pronotum white, body setae brown. Three setae on the U-shaped ridge at the back of the frons. Proximal segment of the antenna with 12 to 13 setae, 2nd segment with 9 to 10 setae. Maxillary palpus with 3 segments. On the abdomen, the median hook has 3 setae, the inner hook with 2 setae. The spine of the inner hook is one third the length of the entire hook (Fig. 102). The 9th abdominal sternum bears two groups of 3 setae each, on the caudal margin.

Larval burrows are between 18 and 43 cm deep (7 to 17 in.) (Hamilton 1925).

Range: Extremely limited. Known only from around the airport in Goose Bay, Labrador (Fig. 103).

Other subspecies of *C. limbata* are known from western North America: *C. l. hyperborea* LeConte in northern Alberta and Saskatchewan; *C. l. nympha* Casey from southern Alberta through Manitoba, North Dakota,

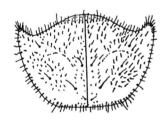

Figure 101. Pronotum of *Cicindela limbata* third instar larva. (Redrawn with permission from Hamilton 1925.)

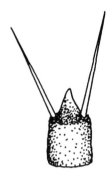

Figure 102. Inner hook of *Cicindela limbata* larva. (Redrawn with permission from Hamilton 1925.)

and into Minnesota; *C. l. limbata* Say from South Dakota, Nebraska, Wyoming, and Colorado; and *C. l. albissima* Rumpp from Utah (Johnson 1989b, Pearson et al. 1997).

Figure 103. Eastern Canadian locality for *Cicindela limbata labradorensis* in Goose Bay, Labrador.

Notes: The origin of *Cicindela limbata labradorensis* is controversial. Johnson (1989b) believes this beetle had a much wider distribution at one time and is now known only from Goose Bay. David Larson (1986) suggested that this beetle may have been introduced to Labrador after 1979, as no specimens were collected prior to that date. He suggests that the current population of this beetle may have originated from individuals transported by aircraft from a population of *Cicindela limbata hyperborea* LeConte, 1863, from western Canada (Larson 1986; Pearson et al. 1997). Visitors to Labrador are encouraged to search for this beetle, for if more localities were found, this would support the idea of a once widely distributed population and subspecies status. The question of the origin of the Goose Bay tiger beetle might also be resolved by biochemical means, using DNA or allozyme techniques to compare western populations with the Labrador population.

Cicindela rufiventris heutzii Dejean, 1831

(Not Illustrated)

Adult Identification: *Cicindela rufiventris heutzii* maculations include a complete middle band, there is a complete to almost complete humeral lunule, and the subapical dot is connected to, or almost connected to, the apical lunule.

Similar Species: Superficially like *Cicindela repanda*, but *C. rufiventris heutzii* has the distinctive red abdomen. *Cicindela rufiventris rufiventris* has maculations reduced to dots (Plate 2).

Habitat: *Cicindela rufiventris heutzii* is localized on granitic rock ledges in eastern Massachusetts; the known localities form an arc around Boston from Gloucester to the Blue Hills (Wilson 1971).

Third Instar Larva Identification: *Cicindela rufiventris heutzii* larvae have not been described, but they presumably burrow in patches of moss and lichens in granitic rock crevasses, making the larvae extremely difficult to secure (Wilson 1971).

Range: Known from around Boston, MA (Fig. 111) and reported from Rhode Island (Bousquet and Larochelle 1993).

Notes: This subspecies is often reported in the literature as *Cicindela rufiventris **hentzi*** or ***hentzii***. However, the original publication by Dejean is *Cicindela rufiventris **heutzii*** (Murray 1980, Bousquet and Larochelle 1993).

(See description of *Cicindela rufiventris rufiventris* for complete account of this species.)

DARK SPECIES WITH MACULATION PATTERN REDUCED

Cicindela hirticollis rhodensis Calder, 1916

Adult Identification: A brownish beetle with maculations varying from almost complete to almost absent. According to Graves et al. (1988), as one proceeds north from Massachusetts along the Atlantic coast, the maculations are more reduced. However, the individuals at one locality may vary from almost fully maculate to having almost no maculations.

Similar Species: Species with reduced maculations that may be confused with *C. hirticollis rhodensis* include *C. punctulata* and *C. longilabris*. *Cicindela punctulata* is found away from water and has a row of green pits along the suture of the elytra (Fig. 13). *Cicindela longilabris* is an upland species and has a very large labrum and a high brow, unlike *C. hirticollis*. *Cicindela hirticollis* can often be distinguished from other species by its very hairy neck.

Habitat: A wet habitat species found on beaches along large bodies of salt or fresh water. Adults are found on the beach, while larvae burrow in the blowing or shifting sand of nearby small dunes. Often found with or near *C. marginata* near salt water.

Range: (Figs. 94 and 95) Individuals with both full and reduced maculations are found along the coast from Massachusetts to Maine (Robert E. Nelson, Colby College, pers. commun.).

Notes: Subspecies have not been found in New Hampshire since 1958, although it was once common (Dunn 1981: 3). Throughout the Great Lakes, *C. hirticollis rhodensis* has disappeared from many of its former localities (Graves et al. 1988).

(See description of *Cicindela hirticollis hirticollis* for complete account of this species.)

Cicindela duodecimguttata Dejean, 1825

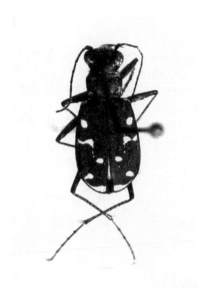

Common Name: 12-spotted tiger beetle.

Adult Identification: 12 to 15 mm. These beetles appear dark brown to black above and metallic greenish-blue underneath. In most specimens, the elytral maculations are broken, forming 6 spots per elytron: the humeral and apical lunules each are represented by 2 round white dots, and the middle band is reduced to a transverse bar and a white knee dot. Very narrow lines may connect some or all of these dots, or in some specimens the dots may be so reduced as to make the beetle appear black. The frons and genae as well as the underside in general are clothed in white setae. The clypeus is glabrous and the epipleura are dark for most of their length. The labrum teeth are either absent entirely (reduced to three wide smooth bumps) or only the middle tooth is present. The setae on the labrum (4 to 5 per side) are set back from the anterior margin about one third of the distance of the length of the labrum. The 1st antennal segment has only 3 erect setae.

Similar Species: Those specimens of *C. duodecimguttata* with the most developed maculations approach those of *C. repanda repanda*, but they never have a well-developed marginal line. However, *C. repanda novas-*

cotiae and *C. duodecimguttata* have very similar maculation patterns. *Cicindela duodecimguttata* has dark epipleura, while *C. repanda* has pale epipleura. Compared to *C. repanda*, the pronotum of *C. duodecimguttata* is shorter, broader, and more flattened above, and the ground color is somewhat darker. At first glance, *C. duodecimguttata* seems almost identical to *C. rufiventris*, viewed from above, but the latter species has a tan-red colored abdomen. *Cicindela longilabris* is another dark species with dots, but it is hairless, with a more strikingly concave "high-brow" frons and a long labrum. *Cicindela punctulata* is also dotted, but is smaller and narrower, the frons is hairless, and the elytra have greenish pits near the scutellum and along the elytral suture (Fig. 13).

Habitat: *Cicindela duodecimguttata* is a wet-habitat species. Found along wooded rivers whose banks are vertical, but on higher shadier and somewhat drier ground than *C. repanda* (Wallis 1961). Also found on wet dark sand of cranberry bog paths and moist logging roads, trails, and paths (Leng 1902; Dunn 1981). The soil the beetles frequent is quite dark, making the beetles difficult to see.

Life History: *Cicindela duodecimguttata* is a spring–fall species, but some individuals live into July. The details of the life cycle have not been reported (Hamilton 1925), but are probably similar to that of *C. repanda* (Shelford 1908). Adults most commonly found from mid-April through June, then again in late July through the beginning of October (Dunn 1981; Larochelle 1986a). Adults seen mating in May and early June in New Hampshire (Dunn 1981).

Third Instar Larva Identification: 18 to 20 mm. The head and pronotum have a coppery-bronze color with green reflections. The setae on the top of the head and pronotum are white, while the other body setae are brown. The pronotum has 10 or fewer secondary setae, and there are no secondary setae along the edges of the meson between primary setae 1 and 5 (Fig. 104). The U-shaped ridge on the back end of the frons has 2 distinct setae. The maxillary palpus has 3 segments. The 1st and 2nd antenna segments each have 9 to 11 setae. On the abdomen, the 9th sternum has two groups of 3 setae on the caudal margin. The median hooks have 3 distinct setae of similar size, while the inner hook has 2 setae on the shoulder (Fig. 105).

Larvae dig burrows at an oblique angle in clay and humus along gently sloping riverbanks. Summer burrows are shallow, 10 to 13 cm (4 to 5 in.) deep, while winter burrows reach 38 cm (15 in.) deep. Larvae will abandon

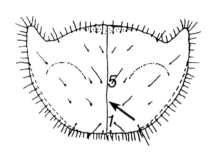

Figure 104. Pronotum of the third instar larva of *Cicindela duodecimguttata.* (Redrawn with permission from Hamilton 1925.) Arrow shows area of meson lacking the pair of secondary setae between primary setae 1 and 5.

Figure 105. Inner hook of *Cicindela duodecimguttata.* (Redrawn with permission from Hamilton 1925.)

their burrows and seek out moist soil to construct new burrows if the soil becomes too dry (Hamilton 1925).

Range: Found throughout our area (Figs. 106, 107) (Bousquet and Larochelle 1993). Occurs west across Canada into central Alberta; north to Lake Athabasca and Great Slave Lake; east to Labrador and Newfoundland (Wallis 1961; Freitag 1965). Widely distributed east of the Rocky Mountains (Graves and Brzoska 1991; Pearson et al. 1997).

Notes: *Cicindela duodecimguttata* is known for its powers of dispersal and is often the first species that colonizes a newly created wet habitat site (Knisley and Schultz 1997).

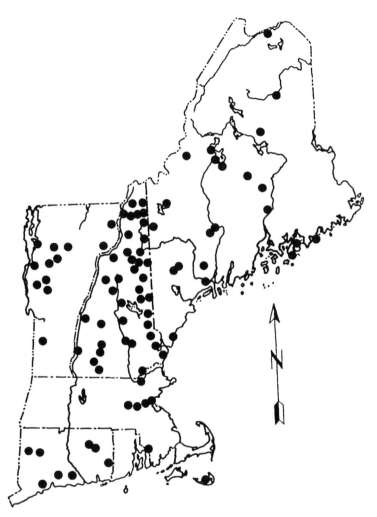

Figure 106. New England localities for *Cicindela duodecimguttata*.

Figure 107. Eastern Canadian localities for *Cicindela duodecimguttata*.

Cicindela repanda novascotiae Vaurie, 1951

(Not Illustrated)

Adult Identification: A brown to light brown beetle with maculations reduced to dots. The basal lunule is usually reduced to dots, or dots just barely connected. The middle band is narrowed or broken into dots. The apical lunule may lack the inward projection.

Similar Species: *Cicindela repanda novascotiae* can be distinguished from *C. r. repanda* by its reduced maculations and lighter color, often being more bronze-reddish than the dark chocolate brown color of *Cicindela repanda repanda*. *Cicindela repanda novascotiae* is most likely to be confused with *Cicindela duodecimguttata*, but this latter species has a wide pronotum, and the epiplura are dark for most of their length, while in *C. repanda* the pronotum is narrow, and the epipleura are pale for most of their length.

Range: *Cicindela repanda novascotiae* is known from Nova Scotia, Prince Edward Island, Magdalen Islands, and Cape Breton Island (Vaurie 1951) (Fig. 71). Bousquet and Larochelle (1993) list *C. repanda novascotiae* from New Brunswick and Quebec.

(See description of *Cicindela repanda repanda* for complete account of this species.)

Cicindela rufiventris rufiventris Dejean, 1825

Common Name: Red-bellied tiger beetle.

Adult Identification: 9 to 12 mm. A <u>dull-black species with reduced and variable maculations</u>. Metallic blue-green head and thorax below with a <u>red-to-tan-yellow abdomen</u>. Humeral lunule reduced to humeral and posthumeral dots; middle band very thin or broken into dots; apical lunule with a lateral bump and isolated subapical dot present. The pronotum is narrow and elytral microserrations are present. An apical tooth is present on the elytron. The long tactile setae on the front and middle trochanters are missing (Fig. 108). The clypeus, frons, and genae are glabrous. The proepisternum is glabrous or with very few setae. The labrum has 6 setae and 1 tooth.

There are two subspecies in our area: *Cicindela rufiventris rufiventris* Dejean, 1825, and *Cicindela rufiventris heutzii* (see page 92.) The latter subspecies has a more complete maculation pattern than the nominate form *C. r. rufiventris*. In *C. r. rufiventris* the humeral lunule is reduced to one or two spots, the marginal band is reduced to 1 or 2 spots, and the middle band is reduced to 2 thin spots (Wilson 1971).

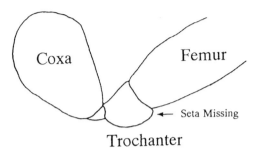

Figure 108. Front leg base of adult *Cicindela rufiventris*. (Redrawn with permission from Willis 1968.)

Similar Species: *Cicindela duodecimguttata* has a very similar maculation pattern, but is shinier, has a wider pronotum, lacks the red abdomen, and is found near water. *Cicindela rufiventris* is a dry-habitat upland species. *Cicindela punctulata* is blackish with reduced maculations, but is small and has a row of green pits or enlarged punctures along the midline on the elytra that are lacking in our other *Cicindela* species.

Habitat: *Cicindela rufiventris rufiventris* is found in eroded areas such as road cuts, embankments, with clay and gravel soil near forest edges. South of our area, commonly found in red-clay hills. Also seen on rock ledges and sunny rock outcroppings and on vegetation covering crushed rock. Often very local and limited in distribution.

Life History: A summer species with a 1-year life cycle (Knisley and Schultz 1997). Adults are common in July and August (earliest record June 17, latest mid-September).

Third Instar Larva Identification: 12 to 17 mm. Labrum and head black with faint purple reflections. Mandibles with black tips and red-brown bases. U-shaped ridge on caudal end of frons with 2 setae. Antennae brown, 1st segment bearing 5 to 6 setae, 2nd segment with 8 to 9 setae. Setae on head and pronotum white, other body setae yellow-brown color. Stemmata 1 and 2 of equal size, with distance between them greater than the diameter of either stemma. Pronotum brown-black color, with green, purple, bronze or copper reflections (Fig. 109). Primary seta number 5 is missing on the pronotum. On the abdomen, the sternum of segment 9

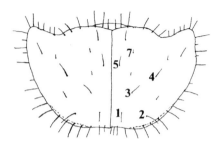

Figure 109. Pronotum of the larva of *Cicindela rufiventris.* Primary seta number 6 is missing (if present it would be just lateral to seta 5); secondary setae are unnumbered. (Redrawn with permission from Beatty and Knisley 1982.)

Figure 110. Inner hook of *Cicindela rufiventris.* (Redrawn with permission from Beatty and Knisley 1982.)

has two groups of 3 long setae on the caudal margin. Median hooks of abdominal segment 5 bearing 3 to 4 setae, inner hooks with 2 to 3 setae on shoulder. Spine of inner hook half the length of the entire hook (Fig. 110) (Beatty and Knisley 1982).

Range: *Cicindela rufiventris rufiventris* is a southern species with *C. r. heutzii* known only from Massachusetts and Rhode Island (Bousquet and Larochelle 1993); while the nominate form (*C. r. rufiventris*) is found in our area from southwestern Vermont (Howard Romack; Cambridge, NY; pers. commun.), western Massachusetts, south through Rhode Island and Connecticut (Fig. 111) (Wilson and Brower 1983; Comboni and Schultz 1989; Bousquet and Larochelle 1993). Intergrades between the two subspecies have been found in Norwell, MA (Fig. 111) (Valenti 1996b). *Cicindela r. rufiventris* is distributed across eastern U.S. from Massachusetts to Mexico (Murray 1980; Pearson et al. 1997).

Notes: *Cicindela rufiventris* has been listed in Connecticut as a "Species of Special Concern" (Sikes 1997).

Adults are not as wary as our other species and will only fly a short distance when disturbed. Often they can be caught without a net, and

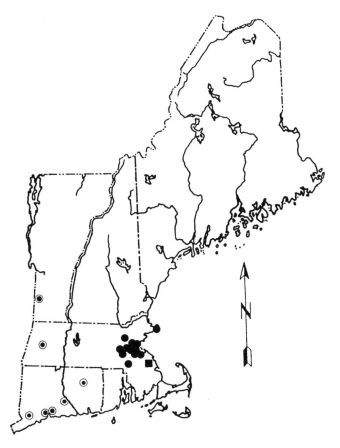

Figure 111. New England localities for *Cicindela rufiventris* and subspecies.
◉ *C. rufiventris rufiventris;* (●) *C. rufiventris heutzii;* (■) intergrades.

are known to come to lights (Graves and Pearson 1973). On very hot days adults avoid direct sun, and are more often seen in the morning or evening, or midday in partial shade.

Cicindela longilabris longilabris Say, 1824

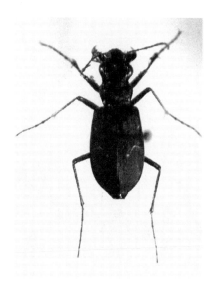

Common Name: Long-lip tiger beetle.

Adult Identification: 15 to 17 mm. A large dark (superficially flat-black colored) species with maculations reduced to very thin lines and tiny dots. The middle band is usually complete but extremely narrow. The marginal line is missing and the lunules are reduced to dots. The frons between the eyes is broadly concave and hairless, giving the insect a "high brow" appearance, and the labrum is much longer and bumpier than in other *Cicindela* species (Fig. 112). The labrum has one distinct middle tooth and lateral bumps. The clypeus and genae are glabrous, as is the frons,

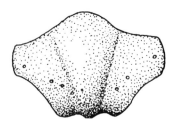

Figure 112. Labrum of adult *Cicindela longilabris*. (Redrawn with permission from Graves 1963.)

except for sensory setae around the eyes. The proepisternum is sparsely covered with setae.

Similar Species: *Cicindela duodecimguttata* is superficially similar, but has a hairy frons, a different labrum, and the middle band consists of 2 to 3 broader obvious dots connected by constrictions (if complete).

Habitat: A northern species found near bogs on sand or fine gravel soil, and on bare rock ledges in the mountains. Also found in sandy areas with mucky dark soil in or adjacent to coniferous forest. Found among jack pine, blueberries, and reindeer moss (*Cladonia*) (Graves 1963). Collected along abandoned or seldom-used sandy roads and other sandy gaps in coniferous forests. Has been collected from above tree line in the Presidential Range of New Hampshire, including the summit of Mt. Washington, 1917 meters (6288 feet) elevation (Wilson and Brower 1983).

Life History: A spring–fall species found from early June into July, then again in mid- to late August.

Third Instar Larva Identification: Leffler (1979) described a second instar *C. longilabris* as follows:

> Head and clypeus rufous with strong brassy and bluish-green metallic reflections; labrum [sic] rufous with brassy reflections only at the base; antennae, basal half of mandibles, maxillae and ventral surface of head light brown, remainder of mandibles dark brown; pronotum rufous with strong cupreous, violet, brassy, and bluish-green reflections; anterior edge and disk of mesonotum dark brown; remainder of mesonotum, metanotum, legs and sclerotized areas of abdomen yellowish-brown; setae of head and pronotum white, other setae brown ... U-shaped ridge on caudal part of frons bearing two setae; ... maxillary palp three-segmented, ... [on the abdomen] inner hooks with fixed spine one-third length of hook and slightly shorter than two setae on shoulder; ventral elevations of sternum IX with three main setae.

We are puzzled by the remark above about the "labrum" being rufous, since tiger beetle (and other adephagous beetle) larvae never have a distinct labrum. We assume Leffler meant frons instead of labrum.

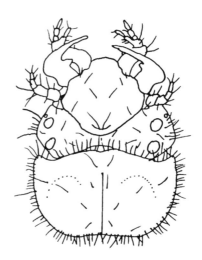

Figure 113.   Head and pronotum of the third instar larva of *Cicindela longilabris*. (Redrawn with permission from Spanton 1988.)

Figure 114.   Inner hook of Hamilton's (1925) species "C," presumably *Cicindela longilabris*. (Redrawn with permission from Hamilton 1925.)

Hamilton (1925, p. 30) described an unidentified *Cicindela* species (species "C") from Pines, IN, which is very likely to be *C. longilabris,* because the larvae of all other species from that region are known.

Spanton (1988) described the third instar larva of *C. longilabris* from larvae reared from captive adults collected near Thunder Bay, ON as follows: There are 2 setae on the U-shaped ridge of the frons. The 1st segment of the antenna has 8 setae, the 2nd segment has 8 or 9 setae, and the 3rd segment has 2 setae. The diameter of stemma 1 and 2 and the distance between them are approximately the same. The pronotum has between 12 and 16 setae (Fig. 113). The median hook bears 2 to 4 setae, and the inner hook has 2 setae that are twice as long as the spine. The spine of the inner hook is approximately one third the length of the hook (Fig. 114).

Although the above descriptions will be helpful in identifying *C. longilabris* larvae, a description of third instar larvae of *C. longilabris* from the northeast needs to be done.

Range: Primarily a northern species (Figs. 115, 116). Found across Canada from near Vancouver to Labrador and Newfoundland. Occurs as far north as Alaska, the Yukon Territory, Great Slave Lake, James Bay, and east to Battle Harbor, Labrador (Wallis 1961, Pearson et al. 1997). In the U.S., found in Maine, New Hampshire, northern New York (Adirondacks), Vermont, Michigan, and extending south in the Rocky Mountains into Montana, Idaho, Wyoming, Colorado, Utah, Arizona, and New Mexico

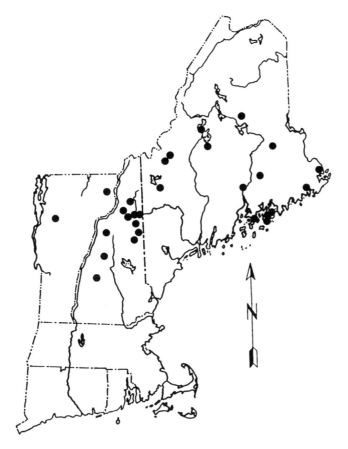

Figure 115.   New England localities for *Cicindela longilabris*.

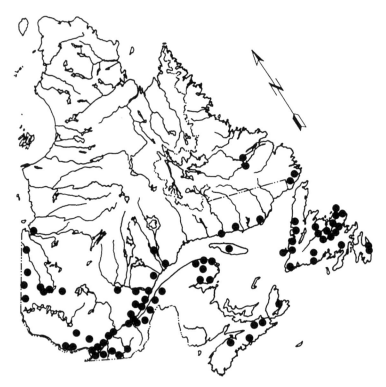

Figure 116. Eastern Canadian localities for *Cicindela longilabris*.

(Bousquet and Larochelle 1993; Pearson et al. 1997). Western varieties are described in Leng (1902).

Notes: *Cicindela longilabris* occurs in many locations, but is rarely common at any one place.

Cicindela punctulata punctulata Olivier, 1790

Common Name: Punctured tiger beetle.

Adult Identification: 11 to 14 mm. A small, slender, black-brownish species with a bare frons, metallic greenish blue underneath and with elytral maculations broken into tiny white dots. This sparsely haired tiger beetle looks leathery and graceful under magnification: they are the "deer" or "gazelle" of northeastern tiger beetles. Each elytron has a row of green foveae (pits or punctures) near the scutellum along the medial margin, a diagnostic feature. Adults emit a fruity odor resembling apples, when handled. Labrum has a single tooth. The clypeus, frons, and genae are glabrous. The 1st antennal segment has only 1 sensory seta.

Similar Species: None: other dark species with broken markings lack the green foveae or pits along the midline. *Cicindela rufiventris* superficially appears similar to *C. punctulata*, but the former species has a reddish abdomen.

Habitat: A generalist found on bare soil, especially in cultivated or human-disturbed land with dry hard-packed soil. The burrows of the larvae are found among the weeds and grasses of this habitat (Graves 1963). Adults are often seen on city sidewalks, dusty dirt roads, and median strips of larger roads in midsummer. Adults have been observed climbing and perching on grass stalks. This is the species you most likely would find in a suburban back yard (Graves and Brzoska 1991).

Life History: A summer species, with most individuals having a 1-year life cycle (Shelford 1908). Hamilton (1925:42) refers to Criddle's (1907) paper, where he suggests that in Manitoba larvae probably take 2 winters before pupating. Adults are found from late June and early July into October and are most common in July.

The female beetle lays eggs in late July in hard dry humus. The larvae reach third instar by fall (first and second instars lasting 3 weeks each). The third instar larva overwinters and begins feeding again the following spring (Shelford 1908; Hamilton 1925). The larvae probably pupate in mid-May through July when adults emerge.

Third Instar Larva Identification: 14 to 16 mm. The head and pronotum are purple-bronze color with faint blue reflections. The setae on the top of the head and pronotum (Fig. 117) are white, other body setae are brown. The diameter of stemma 2 is less than the distance between stemmata 1 and 2. The 1st antennal segment has 5 to 6 setae, while the 2nd segment has 9 or 10. The U-shaped ridge on the back of the frons has 2 distinct setae. On the abdomen, the sternum of segment 9 has two groups of 3 setae on the caudal margin. The median hook has 3 distinct setae of similar size; the inner hook has 2 setae on the shoulder (Fig. 118).

Burrows are found on dry hard-packed soil between clumps of moss or grass and range from 30 to 40 cm deep (12 to 26 in.). The pupal chamber is constructed about halfway down the burrow (Fig. 119).

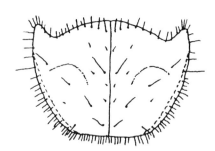

Figure 117. Pronotum of the larva of *Cicindela punctulata*. (Redrawn with permission from Hamilton 1925.)

Figure 118. **Inner hook of** *Cicindela punctulata.* **(Redrawn with permission from Hamilton 1925.)**

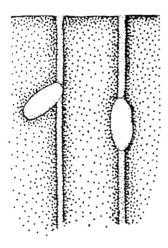

Figure 119. **Burrows of** *Cicindela punctulata* **showing variation in orientation of pupal chambers. (Redrawn with permission from Shelford 1908.)**

Range: Reported from all New England States, Quebec, and New Brunswick (Figs. 120, 121) (Bousquet and Larochelle 1993). Collected from southern Canada as far west as Alberta through Ontario. Widely distributed from Montana south to Mexico and east to Florida (Pearson et al. 1997).

Notes: Adults attracted to lights at night. One of the most common and smallest species in the northeast.

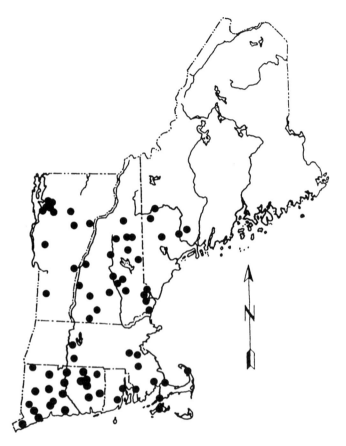

Figure 120. New England localities for *Cicindela punctulata*.

Figure 121. Eastern Canadian localities for *Cicindela punctulata.*

PLATE 1

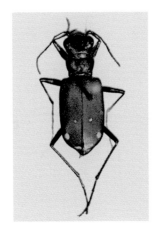

**Six-Spotted
Tiger Beetle**
Cicindela sexguttata
Fabricius, 1775.

Tiger Beetle
*Cicindela patruela
patruela* Dejean, 1825.

**Common Shore
Tiger Beetle**
*Cicindela repanda
repanda* Dejean, 1825.

Big Sand Tiger Beetle
*Cicindela formosa
generosa* Dejean, 1831.

**White Mountain
Tiger Beetle**
Cicindela ancocisconensis
T.W. Harris, 1852.

**Oblique-Lined
Tiger Beetle**
*Cicindela tranquebarica
tranquebarica* Herbst,
1806.

PLATE 2

Puritan Tiger Beetle
Cicindela puritana
G.H. Horn, 1871.

**Hairy-Necked
Tiger Beetle**
*Cicindela hirticollis
hirticollis* Say, 1817,
full maculations.

**Hairy-Necked
Tiger Beetle**
*Cicindela hirticollis
rhodensis* Calder, 1916,
reduced maculations,
genitalia exposed.

**Salt Marsh
Tiger Beetle**
Cicindela marginata
Fabricius, 1775, female.

**12-Spotted
Tiger Beetle**
Cicindela duodecimguttata
Dejean, 1825.

**Red-Bellied
Tiger Beetle**
*Cicindela rufiventris
rufiventris* Dejean,
1825.

PLATE 3

**Long Lip
Tiger Beetle**
*Cicindela longilabris
longilabris* Say, 1824.

**Punctured
Tiger Beetle**
*Cicindela punctulata
punctulata* Olivier,
1790.

Cow Path Tiger Beetle
*Cicindela purpurea
purpurea* Olivier, 1790.

Clay Bank Tiger Beetle
Cicindela limbalis
Klug, 1834, red form.

Clay Bank Tiger Beetle
Cicindela limbalis
Klug, 1834, green form.

Smooth Tiger Beetle
*Cicindela scutellaris
lecontei* Haldeman,
1853.

PLATE 4

**Cobblestone
Tiger Beetle**
Cicindela marginipennis
Dejean, 1831.

**Northeastern Beach
Tiger Beetle**
*Cicindela dorsalis
dorsalis* Say, 1817.

Ghost Tiger Beetle
Cicindela lepida
Dejean, 1831.

Carolina Tiger Beetle
*Tetracha carolina
carolina* (Linne, 1767),
ovipositor exposed.

S-Banded Tiger Beetle
*Cicindela trifasciata
ascendens* LeConte,
1851.

Goose Bay Tiger Beetle
*Cicindela limbata
labradorensis* Johnson,
1990.

BRONZED SPECIES WITH REDUCED MACULATIONS AND PROMINENT MIDDLE BANDS

Cicindela limbalis Klug, 1834

Red form   **Green form**

Common Name: Clay bank tiger beetle.

Adult Identification: 14 to 16 mm. This species and the next species, *C. purpurea*, are very similar, both with metallic color varying from bright red-purple-bronze to partially or wholly greenish above and metallic greenish-blue below. The most common coloration above is red-purple-bronze in the middle of the body with green to bluish color along the lateral margins and on the front and back margins of the pronotum. The lateral margin of the elytra appears as a dark indigo-blue black band. *Cicindela limbalis* has a long middle band with a distinct knee and foot. The middle band also has a long outer segment that reaches the dark lateral band. The posthumeral dot is large. The clypeus is glabrous, while the frons, genae, and lateral third of the pronotum are covered with erect setae. The labrum has 3 teeth. The first antennal segment has approximately 8 setae.

Similar Species: *Cicindela purpurea purpurea* has the posthumeral dot reduced or absent, and the middle band is a short oblique fragment with no obvious knee and foot compared to that in *C. limbalis*.

Habitat: Found on bare <u>sloping</u> (often steep) clay soil, as on abandoned hill farms and road cuts. Recorded only from hilly regions. Larvae found at the bottom of steep slopes of sparsely vegetated clay banks and hillsides, while adults are found on the banks and the flat areas adjacent. Sometimes found on the clay banks along rivers.

Life History: A spring–fall species with a 3-year life cycle (Schincariol and Freitag 1991). Adults found from mid-April through June, then again from late August to mid-September. Mating pairs seen in New Hampshire on May 11 (Dunn 1981). Eggs laid in June; first instars are present from June through early August. Larvae hibernate through the first winter as second-instar larvae. They molt into third instars the following year in June, pupate in July, and emerge as adults (and go above ground briefly if at all) in August. Other newly emerged adults remain in their pupal chambers for another winter, while the other adults rebuild a hibernaculum burrow after a brief time above ground (Hamilton 1925; Knisley and Schultz 1997).

Third Instar Larva Identification: 19 to 22 mm. Head and pronotum purplish-bronze with metallic green reflections. Setae on top of head and pronotum (Fig. 122) white, other setae brown. The 1st antennal segment has 5 or 6 setae, the 2nd segment with 7 or 8 setae. On the abdomen, median hooks have 2 setae (Fig. 123). Inner hooks have 2 setae and the central spine of the inner hook projects more than one half the length of the entire hook (Fig. 124). Very similar to *C. purpurea*, but with many more secondary setae on the pronotum, and a longer projecting spine on the inner hook (Hamilton 1925).

Larvae burrow into steeply sloping clay soil (Fig. 125). The burrow entrance tunnel is usually perpendicular to the face of the clay bank; then

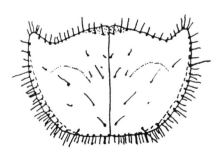

Figure 122. Pronotum of the larva of *Cicindela limbalis*. (Redrawn with permission from Hamilton 1925.)

Median Hooks

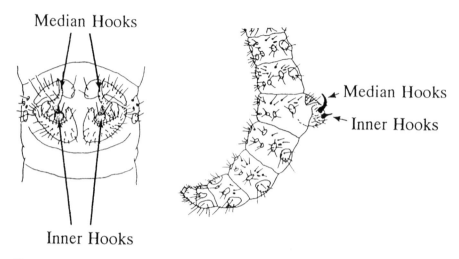

Median Hooks

Inner Hooks

Inner Hooks

Figure 123. Dorsal and lateral view of median and inner hooks of third instar larva of *Cicindela limbalis*. (Redrawn with permission from Hamilton 1925.)

Figure 124. Inner hooks of *Cicindela limbalis*. (Redrawn with permission from Hamilton 1925.)

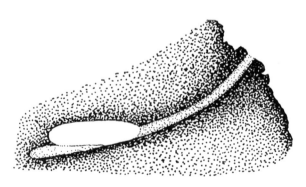

Figure 125. Burrow of *Cicindela limbalis*, showing the orientation of the pupal chamber. (Redrawn with permission from Shelford 1908.)

the burrow becomes horizontal near its end, approximately 7 to 10 cm (3 to 4 in.) deep. A chimney-shaped structure is often formed by the excavating larva at the burrow's mouth (Hamilton 1925). The pupal chamber is made from enlarging the rear of the burrow.

Range: Found throughout our area except Labrador (Figs. 126, 127). Common on Mt. Desert Island, ME (Wilson and Brower 1983). Northwest to Fort Norman in Canada's Northwest Territories near Great Bear Lake. East from Alberta through south central Canada to western Newfoundland (Wallis 1961). Throughout the northeast, south to Pennsylvania, west to Utah and New Mexico (Pearson et al. 1997).

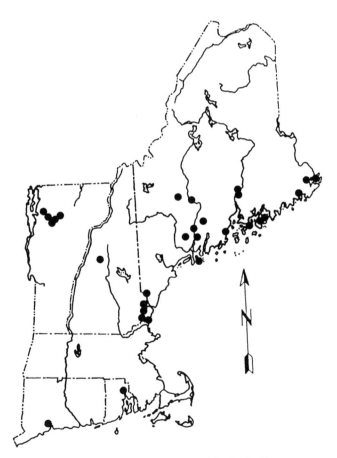

Figure 126. New England localities for *Cicindela limbalis.*

Figure 127. Eastern Canadian localities for *Cicindela limbalis*.

Cicindela purpurea purpurea Olivier, 1790

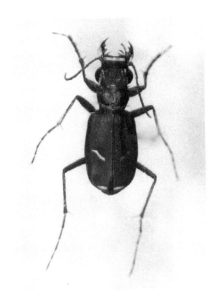

Common Name: Cow path tiger beetle.

Adult Identification: 14 to 16 mm. This species and *C. limbalis* are very similar, both with metallic color varying from bright red-purple-bronze to partially or wholly greenish. The most common coloration above is red-purple-bronze in the middle of the body with green to bluish color along the lateral margins and on the front and back margins of the pronotum. Underneath, this species is bright cupreous red to purple on the head and thorax, and metallic greenish-blue on the abdomen. The sides of the head, the base and apex of the pronotum, and a narrow marginal stripe on the elytra are contrastingly dark green (blue-black on the greenest specimens). The white maculations are reduced to dots and there is no trace of the white marginal line. The posthumeral dot in *C. purpurea* is very small or absent, and the middle band is short and oblique. As in *C. limbalis*, the clypeus is glabrous while the frons, genae, and lateral thirds of the pronotum are covered with erect setae. The labrum has 3 teeth and the 1st antennal segment has approximately 8 setae.

Similar Species: *Cicindela limbalis* has a large posthumeral dot and the middle band is more developed than in *C. p. purpurea*, with a longer transverse outer segment that reaches the dark lateral band. *Cicindela* *sexguttata* resembles the greenest individuals, but has a hairless frons and no trace of bronzing.

Habitat: An upland species found over our general area on bare <u>level</u> clay/loam soil. Often found in meadow paths, grassy roadsides, grassy areas of sand dunes, and forest clearings. Found on low spots covered with dark blackened decaying vegetation where water pools after rains.

Life History: An early spring–fall species with at least a 2-year life cycle. Records from New England include adults found as early as March 3 (New Hampshire) through mid-June, then again in late August through mid-October (Larochelle 1986a, Dunn 1981). In Maine, adults have been collected in the first days of May (Wilson and Brower 1983). In Ohio, adults have been collected in late April through June, and again in September and October (Graves and Brzoska 1991). In New Jersey, they are found from early April to the first week in May, then again in September (Boyd 1978).

Adults mate and lay eggs in the spring in grassy places where partially moist dark soil is exposed such as along trails, streambeds, or abandoned dirt roads. Larvae go into hibernation as third instars in the fall. Larvae continue to feed the following spring, and pupate in late June and July, then emerge as adults in August. Adults create a hibernaculum burrow for another winter and emerge in spring sexually mature (Hamilton 1925).

Third Instar Larva Identification: The following description from Hamilton (1925) is of *Cicindela purpurea* var. *graminea* Schaupp, which is now recognized as a synonym for *C. p. audubonii* LeConte, 1845 (Bousquet and Larochelle 1993).

19 to 22 mm. Head and pronotum purplish-bronze with metallic green reflections. Setae on top of head and pronotum white, other setae brown. The U-shaped ridge on the caudal part of the frons has 2 setae. The secondary setae on the pronotum are missing, except for 2, one of which is a large seta found cephalo-laterad of seta 4 (Fig. 128). The first antennal segment has 6 or 7 setae, the second segment has 8 or 9 setae. On the abdomen, median and inner hooks each have 2 setae. The central spine of the inner hook projects one half or less the entire length of the hook (Fig. 129). The sternum of the 9th abdominal segment has two groups of 4 setae along the caudal margin. In general, this larva is very similar to *C. limbalis*, except with few secondary setae on the pronotum, and the central spine of the inner hook does not project as far (Hamilton 1925).

Burrow descends straight down with pupal chamber constructed near the bottom (Fig. 130).

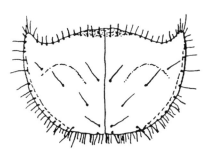

Figure 128. Pronotum of the larva of *Cicindela purpurea audubonii*. (Redrawn with permission from Hamilton 1925.)

Figure 129. Inner hook of *Cicindela purpurea audubonii*. (Redrawn with permission from Hamilton 1925.)

Figure 130. Burrow of *Cicindela purpurea* showing the orientation of the pupal chamber. (Redrawn with permission from Shelford 1908.)

Range: Reported throughout our area except in Labrador, Newfoundland, and New Brunswick (Figs. 131, 132) (Bousquet and Larochelle 1993). Widely distributed in North America. Found from Vancouver to Quebec, in most of the U.S., except perhaps the southernmost states of Louisiana and Florida (Leng 1902, Pearson et al. 1997).

Notes: A harbinger of spring, *C. purpurea* is not often collected.

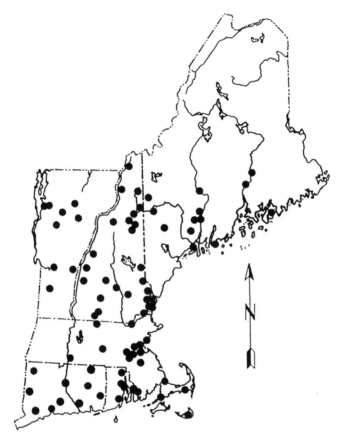

Figure 131. New England localities for *Cicindela purpurea.*

Figure 132. Eastern Canadian localities for *Cicindela purpurea.*

SPECIES WITH DISTINCTIVE MARGINAL BANDS

Cicindela scutellaris lecontei Haldeman, 1853

Common Name: Smooth tiger beetle.

Adult Identification: 11 to 13 mm. An attractive red-purplish-bronze species with smooth elytra. Metallic greenish-purple below. The <u>marginal band is reduced and points toward the midline, making a broad triangle.</u> The middle band is missing, while the apical lunule is usually intact and the humeral lunule may be divided into 2 dots, the posterior dot sometimes joined to the marginal line. The frons is hairy and the labrum is white in the male, while in the female the frons has only a few hairs near the eyes and the labrum is largely black. The labrum has 3 teeth, and the first antennal segment has between 15 and 20 large white setae. The pronotum has 2 longitudinal lines of 5 to 8 setae, with each line located at about two thirds the distance from the midline to the lateral edge of the pronotum.

Cicindela scutellaris is an extremely variable species. A green subspecies *Cicindela scutellaris rugifrons* is known. Bousquet and Larochelle (1993) list seven recognized subspecies in the U.S. (not including the "modesta" color form). According to Wallis (1961), *C. scutellaris* is the most variable species in all of North America. A thorough study of the variation and subspecies of *C. scutellaris* needs to be done.

Similar Species: *Cicindela scutellaris* is not easily confused with other species. *Cicindela limbalis* and *C. purpurea* are mixed bronze and green, but these species have a distinct middle band, and the triangular marginal band of *C. scutellaris* is missing. *Cicindela s. lecontei* is red color, while a green subspecies, *C. scutellaris rugifrons* is known. Intergrades of these subspecies have been collected (Davis 1903, Wilson and Brower 1983, Dunn 1986).

Habitat: *Cicindela scutellaris* is a dry-habitat species. It is found in sand blowouts, loose sand with sparse vegetation such as grass, lichens, or semiopen areas crossed with trails (Boyd 1978). Also found in bare loose sand dunes, road cuts, and sand pits. Often found with *C. formosa generosa*.

Life History: A spring–fall species with a 2-year life cycle. Adults appear from mid-April through June, then again in September. Mating has been observed on May 14 to June 4 (Dunn 1981).

Adult females lay eggs in dry sand with small amounts of humus in late spring. Larvae go into hibernation as third instars in the fall. Larvae continue to feed the following spring, and pupate in late June and July, then emerge as sexually immature adults in August. Adults create a hibernaculum burrow for another winter and emerge in spring sexually mature (Hamilton 1925).

Third Instar Larva Identification: 20 to 24 mm. The head and pronotum are purple-bronze color with metallic green reflections. The setae on the top of the head and pronotum (Fig. 133) are white, while the rest of the body setae are brown. The U-shaped ridge at the back of the frons has 3 setae. The first antennal segment has 5 or 6 setae, while the second

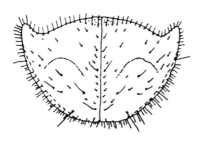

Figure 133. Pronotum of the larva of *Cicindela scutellaris*. (Redrawn with permission from Hamilton 1925.)

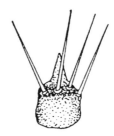

Figure 134. Inner hook of *Cicindela scutellaris*. (Redrawn with permission from Hamilton 1925.)

segment has 10 to 12. On the abdomen, the median hook has 3 setae, while the inner hook has (most commonly) 4, or rarely 3, setae. The inner hook's projecting spine is about one third the length of the entire hook (Fig. 134).

Burrows are 15 to 55 cm (6 to 22 in.) deep, and are often found near those of *C. formosa generosa,* but further away from open shifting sand and more under the protection of vegetation and tree canopy cover (Fig. 135) (Shelford 1908, Hamilton 1925, Knisley and Schultz 1997).

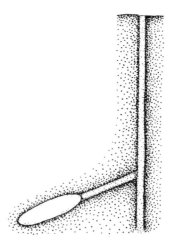

Figure 135. Burrow of *Cicindela scutellaris* showing the orientation of the pupal chamber. (Redrawn with permission from Shelford 1908.)

Range: In our area found from Connecticut, Massachusetts, Maine, New Hampshire, Quebec, and Vermont (Figs. 136, 137). In Canada, found in southern Quebec, Ontario, and west to Manitoba (Wallis 1961, Nelson and LaBonte 1989). *Cicindela scutellaris* in the broad sense is widely

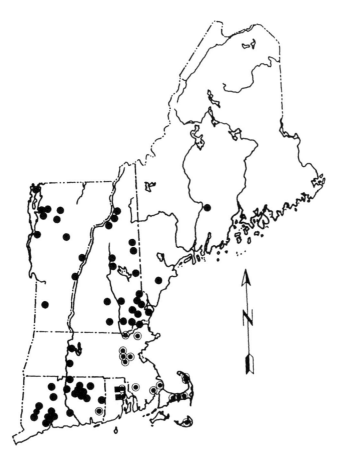

Figure 136. New England localities for *Cicindela scutellaris*. (●) *C. scutellaris*; (◉) *C. scutellaris rugifrons*; (■) black form "modesta."

distributed east of the Rocky Mountains (Pearson et al. 1997). *Cicindela s. lecontei* is found in the U.S. from the uplands of New England (southern Maine, New Hampshire, Vermont, western Massachusetts, Connecticut) south to Tennessee and west to Wisconsin (Dunn 1981). *Cicindela scutellaris rugifrons* and is found in eastern Massachusetts, Rhode Island, Connecticut, and Atlantic coastal states, south to North Carolina (Bousquet and Larochelle 1993).

Notes: *Cicindela scutellaris* is a gregarious species. When a net is put over this beetle, it will often try to hide by burying itself in the sand. Often they may be caught by hand or easily photographed at close range.

Figure 137. Eastern Canadian localities for *Cicindela scutellaris lecontei*.

Shelford (1908) reports that the larvae of the bee fly (Bombyliidae) *Spogostylum anale* Say parasitize about 7% of *C. scutellaris lecontei* larvae.

Cicindela marginipennis Dejean, 1831

Common Name: Cobblestone tiger beetle.

Adult Identification: 12 to 14 mm. A long, lanky olive-green species with a broad white margin on the elytra, and without other white maculations. The marginal line is complete and bulges slightly, showing the remains of the humeral lunule, the middle band, and the apical lunule. The abdomen is bright red and exposed during flight. Trochanters without long tactile subapical setae. The clypeus, frons, and genae are glabrous. The labrum has a single small tooth. Antennal segment one with only a single seta. The coxae and femurs of the front and middle legs are covered with white decumbent hairs.

Similar Species: None.

Habitat: An extremely restricted habitat: found on the beaches of vegetated islands in big rivers and on the cobblestone deltas at the confluence of rivers. The upstream ends of the beaches of the islands are covered with large cobblestones, the downstream ends with sandbars. The beetles are most commonly found on the beach on the sides of these islands (Dunn and Wilson 1979, Dunn 1981). Adults are also common on the cobblestone bars, where they are very difficult to net because of the irregular ground surface. Adults may be found on the riverbanks opposite the islands. On a hot day, the cobblestones radiate heat like stove burners; the beetles stand as high as possible on their long legs to escape the heat. The dense layer of white hairs on their bellies also protects them.

This species may possibly live in similar habitats on many other rivers throughout our area, although dam construction and pollution have eliminated many likely sites.

Life History: A summer species: collected from June through September. Newly emerged beetles were observed on July 13 on the Winooski River in Vermont, and on July 17 on the Connecticut River in New Hampshire (Dunn and Wilson 1979, Dunn 1981). Mating pairs were seen on July 17 and August 7 (Dunn 1979).

Third Instar Larva Identification: The larva is undescribed. Larvae of (presumed) *C. marginipennis* were taken by Wilson and Dunn near the interior of the islands (Dunn 1981: 6), where their burrows were found between cobblestones.

Range: In our area, known only from the Winooski River in the Champlain Basin (found in 1997 by the author), and the Connecticut River and its tributaries in Vermont (Fig. 138). Found associated with large rivers from New England west to the Mississippi (Dunn and Wilson 1979, Graves and Pearson 1973). Found in Alabama, Indiana, Mississippi, New Jersey (islands on the upper Delaware River), New York, Ohio, Pennsylvania, and West Virginia (Bousquet and Larochelle 1993, Pearson et al. 1997).

Notes: This species is threatened with extinction and should not be collected. Along the Connecticut River only five sites are known where small populations exist as of 1989. *Cicindela marginipennis* is listed as a "Rare, Threatened, and Endangered" species in Vermont, and is currently under federal review to be added to the threatened or endangered national list. (For more information contact the Vermont Natural Heritage Program: Dept. of Fish and Wildlife, 103 South Main Street, Waterbury, VT 05676.)

A wary species that flies when disturbed.

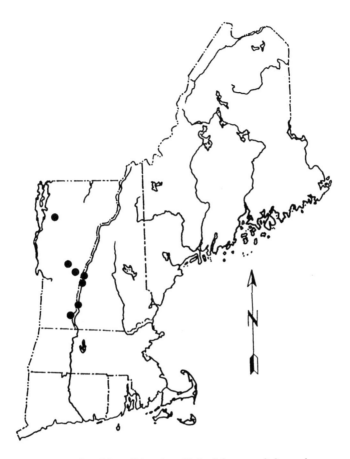

Figure 138. New England localities for *Cicindela marginipennis*.

SPECIES WITH PALE ELYTRA

Cicindela dorsalis dorsalis Say, 1817

Common Name: Northeastern beach tiger beetle.

Adult Identification: 13 to 15 mm. A striking tiger beetle, not likely to be confused with any other species. White elytra with maculations expanded so they join into one another and leave three dark bronzed background curves on each elytron. Head and thorax greenish-bronzed. Underneath bronze with the sides thickly covered with many dense decumbent white hairs. Some of these hairs wrap up and around the beetle's pronotum. The clypeus, frons, and genae are glabrous. The elytra appear to be shaped like partially pointed ovals in females and rounded ovals in males (Fig. 139). Legs are very long. The labrum is wide with one tooth. Males have a long ventral-pointing tooth on the mandible (Fig. 140).

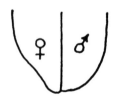

Figure 139. Shape of the female and male elytra of *Cicindela dorsalis*. (Redrawn with permission from Leng 1902.)

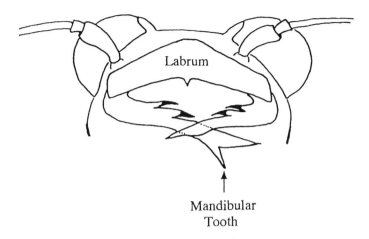

Figure 140. Labrum and mandibular tooth of adult male *Cicindela dorsalis dorsalis.*

Similar Species: The only species possibly confused with *C. dorsalis* is the little gray ghost of the dunes, *C. lepida*, but *C. lepida* is smaller, is found on dunes away from water, and has a more rounded shape, and short legs.

Habitat: Saltwater white-sand beaches with dunes or cliffs on the upper beach. Larvae may burrow from the high-tide mark up to the base of the dunes. This species has been found in the same habitat as *C. hirticollis*. In the past, a population of this species was known from the shores of Lake Michigan (Hamilton 1925: 21).

Life History: A summer species. In southern New England, adult populations reach their peak numbers around the 20th of July (Philip Nothnagle, pers. commun.). Adults were once found in "great swarms" in July (Leng 1902). Now beetles are extremely rare.

Most larvae take 2 years to mature and migrate up the beach in October to avoid winter storm damage. First instar larvae appear in late July and early August, second instars by mid-September. Larvae overwinter as late second instars or early third instars in their 1st year, and as third instars in their 2nd year. (U.S. Fish and Wildlife Service 1993b; Philip Nothnagle pers. commun.).

Third Instar Larva Identification: Larvae of the northeastern beach tiger beetle, *Cicindela dorsalis dorsalis* are undescribed. However, Hamilton (1925) described *Cicindela dorsalis saulcyi* as follows:

15 to 17 mm. There are strong blue reflections from the coppery bronze-colored head and pronotum. Setae on the top of the head and pronotum are transparent, while other body setae are brown. The maxillary palpus is 2-segmented. The U-shaped ridge on the caudal part of the frons has 4 to 6 setae. The pronotum has more than 150 secondary setae (Fig. 141). The 1st segment of the antenna (which is shorter than the second segment) has 5 or 6 setae; the 2nd segment has 9 or 10 setae. The 9th abdominal sternum's caudal margin has two groups of 3 setae. The median hook has 2 setae; the inner hook has 6 or 7 setae. Inner hook spine projects only one fifth the length of the entire hook (Fig. 142) (Hamilton 1925).

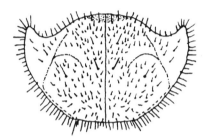

Figure 141. Pronotum of the third instar larva of *Cicindela dorsalis saulcyi*. (Redrawn with permission from Hamilton 1925.)

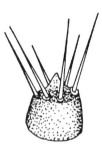

Figure 142. Inner hook of *Cicindela dorsalis saulcyi*. (Redrawn with permission from Hamilton 1925.)

Burrows are between 30 and 46 cm (12 to 18 in.) deep (Hamilton 1925). Barry Knisley has observed *C. dorsalis* larvae crawling away from the water to establish new larval burrows farther up the beach (Knisley and Schultz 1997). Tiny amphipods that live between the intertidal zone to above the strand line are the primary food source of the larvae. Larval densities of 8 larvae per square meter have been observed in the intertidal zone (Knisley and Schultz 1997).

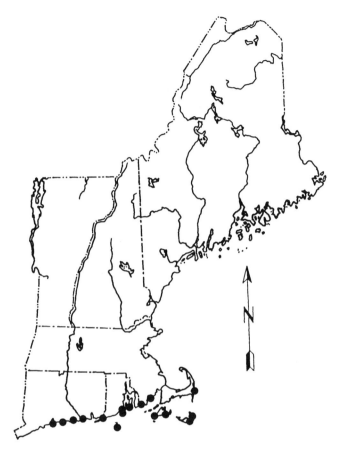

Figure 143. New England localities for *Cicindela dorsalis dorsalis*.

Range: Historical records from our area are known from Cape Cod south along the coast (Fig. 143). Years ago known from the coasts of Connecticut, Delaware, Massachusetts, Maryland, New Jersey, New York, Pennsylvania, Rhode Island, and Virginia. Only a few populations are left on remote beaches in Virginia, Maryland, New Jersey, and two sites in Massachusetts. A variant population is now isolated at Calvert Cliffs, Maryland (Boyd 1975, Boyd and Rust 1982). The smaller subspecies, *C. dorsalis media* LeConte, 1857, is known along the Atlantic coast from southern New Jersey to Florida (Boyd 1978, Bousquet and Larochelle 1993, Yarbrough and Knisley 1994). The two other subspecies are *C. dorsalis saulcyi* Guérin-Méneville, 1840, which ranges from the Florida gulf coast to Mississippi, and *C. dorsalis venusta* LaFerté-Sénectère, 1841, which is found from Louisiana south to Mexico (Graves and Pearson 1973, Boyd and Rust 1982, Knisley et al. 1987, Bousquet and Larochelle 1993, Pearson et al. 1997).

Notes: This species is endangered with extinction in Massachusetts, and listed as threatened with the U.S. Fish and Wildlife Service. The cause of the dramatic decline in numbers during this century is beach development and construction of seawalls, breakwaters, and other stabilization structures; and increased foot traffic, and beach vehicles that crush adult beetles and larvae and compact the beach sand (Stamatov 1972; Knisley et al. 1987; Schultz 1988; Philip Nothnagle pers. commun.).

Marked individuals of *Cicindela dorsalis dorsalis* have been known to disperse 15 miles from the original marking site (Knisley and Schultz 1997).

The U.S. Fish and Wildlife Department and the Nature Conservancy are trying to reintroduce *C. dorsalis* into suitable beaches in the northeast (U.S. Fish and Wildlife Service 1993b).

Any sightings of this beetle in New England should be reported to the Nature Conservancy in your state, the U.S. Fish and Wildlife Service, and the authors.

Robber flies (Asilidae) and wolf spiders (Araneae: Lycosidae) have been important predators of adult *C. dorsalis*. The tiphiid wasp *Methocha* is an important parasitoid of second and third instar larvae (Knisley and Schultz 1997).

Cicindela lepida Dejean, 1831

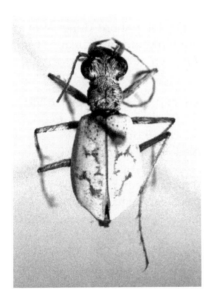

Common Name: Dune ghost tiger beetle.

Adult Identification: 8 to 12 mm. A small round-oval sand-colored tiger beetle with a hairy gray-greenish to reddish-bronze head and thorax. The maculations are expanded to almost completely cover the elytra, leaving only light brown traces. The beetle's underside is brownish green bronzed and covered with dense white decumbent hairs. These hairs cover the entire body except for the eyes, mandibles, labrum, and elytra. The legs and antennae vary between light brick color to the color of sand. The apical border of the elytron is sinuate (especially in females). The labrum is wide with 1 tooth and has 12 long unpigmented setae and a few decumbent white setae.

Similar Species: Only *Cicindela dorsalis dorsalis* comes close to resembling the little ghost. The northeastern beach tiger beetle however, is larger, longer, and restricted to beaches on the Atlantic coast.

Habitat: A dry-habitat species found on deep pure white to pale yellow sloping dune sand away from vegetation. Known as "the little gray ghost of the dunes" (Boyd 1978) the beetle is almost invisible until it moves, because of its cryptic coloration, and like the ghost crab of southern beaches, during the day this beetle is less visible than its own shadow. *Cicindela lepida* is a dune specialist found in undisturbed deep sand on seaward coastal and large lake shore dunes, but is also found on sand flats away from water and in dunes well inland (Gaumer 1970). Sometimes found along with *C. formosa generosa* and *C. scutellaris lecontei*.

Life History: A summer species with a 2-year life cycle. Adults are found from late June through September, most commonly in July.

Shelford (1908) and Hamilton (1925) describe the life history of this species as unique from all other *Cicindela* because the adults live for only about a month, while the larvae live almost 2 years. Two broods, 1 year apart, often are found at the same locality.

Mating sometimes occurs in shallow burrows dug in the sand. Adult females lay yellow cream colored eggs in midsummer (late July in Illinois) and the larvae prepare to hibernate as second instars in the fall. The following spring and summer the larva feeds and reaches third instar by early summer. The third instar continues to feed, hibernates again during the following winter, emerges in the spring, and pupates in June or July of the 2nd year. Adults emerge in midsummer, mate, lay eggs, and die.

Third Instar Larva Identification: 14 to 16 mm. The head and prono-tum are bronze color with metallic green-blue reflections. The setae on head and pronotum (Fig. 144) are long, thin, and appear transparent. Two obvious setae are on the U-shaped ridge at the back of the frons. The proximal segment of the antenna has 6 or 7 setae, while the 2nd segment has 9 or 10 setae. The maxillary palpus has 3 segments. On the abdomen, both the median hook and the inner hook (Fig. 145) have 2 setae. The

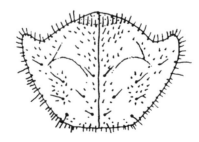

Figure 144.   Pronotum of the larva of *Cicindela lepida*. (Redrawn with permission from Hamilton 1925.)

Figure 145. Inner hook of *Cicindela lepida*. (Redrawn with permission from Hamilton 1925.)

sternum of the 9th abdominal segment has two groups of 3 setae along the caudal margin. The spine on the inner hook is only one sixth the length of the entire hook.

Burrows are constructed in slightly shifting sand or on the tops of dunes. Burrows are narrow but deep, ranging from 60 to 90 cm (25 to 72 in.)! A funnel often forms at the burrow's mouth (Fig. 146). Pupal chambers are long and curved and constructed a few centimeters below ground off the main burrow tunnel (Shelford 1908, Hamilton 1925).

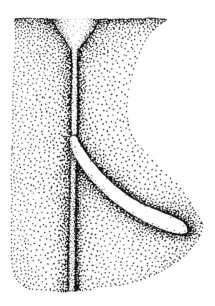

Figure 146. Burrow of *Cicindela lepida* showing the orientation of the pupal chamber. (Redrawn with permission from Shelford 1908.)

Figure 147. Eastern Canadian locality for *Cicindela lepida*.

Range: In our area, reported in Boyd's 1982 checklist, from Massachusetts but the locality information is unknown (Howard P. Boyd, pers. commun.). In Quebec (Fig. 147), known from a single site (Belanger 1982; Bousquet and Larochelle 1993). Recorded from just south of New England on Long Island, NY (Lee H. Herman, pers. commun.; Leonard 1928).

In Canada, also found near Lake Ontario and westward into southern Saskatchewan. Widely distributed in North America east of the Rocky Mountains (Pearson et al. 1997).

Notes: This species is a rare treat and is indicative of undisturbed dune habitat (now very rare). Any sightings or collections of this species should be reported to your local conservation organization (such as The Nature Conservancy), to governmental environmental offices (state, provincial, or federal, such as Agriculture Canada or the U.S. Department of Interior Fish and Wildlife Service) and to the authors.

Adults are not very wary and only fly a short distance when disturbed. The little ghost avoids the heat of midday and is found most often in the morning and evening hours or after dark. It has been attracted to lights at night (Vaurie 1950).

Cicindela punctulata and *C. lepida* are the smallest *Cicindela* species in our area.

STRAY SPECIES AND QUESTIONABLE RECORDS

Tetracha carolina carolina (Linné, 1767)
(= *Megacephala carolina,* see Huber 1994)

Common Name: Carolina tiger beetle.

Adult Identification: 19 to 22 mm. An impressively <u>large metallic green</u> <u>tiger beetle with a large cream-colored apical lunule on the apex of each</u> <u>elytron</u>. Jaws, frons, antennae, and legs a light tan-yellow color. <u>Elytra</u> <u>covered with obvious punctures and with a red-purple iridescence near</u> <u>the midline</u>. The pronotum is very wide at the anterior end, but narrows considerably posterior. The body is glabrous.

Similar Species: None.

Habitat: A mesic forest species found during the day under cover of rocks and logs.

Life History: Unknown.

Third Instar Larva Identification: 25 to 30 mm. The head and pronotum are a deep purple color with metallic green reflections. Caudal margin of pronotum pearly white. Setae on the head and pronotum are brown or occasionally white. Median hooks of the 5th abdominal segment are thorn-like and almost straight (Fig. 148) (this character separates *Tetracha* from *Cicindela* larvae). Inner hooks are small versions of the median hooks (about half the size). The palpiger is membranous (Fig. 149) and the proximal segment of the labial palpus lacks spine-like projections found in *Cicindela* on the ventro-distal margin (Hamilton 1925).

Burrows have been discovered in a variety of soils and habitats including disturbed soil, moist sand on beaches, clay soil, gravel, and moist black loam. Burrows go straight down into the soil, are large, and are between 20 to 30 cm (8 to 12 in.) deep.

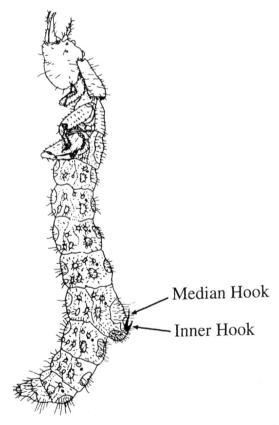

Median Hook

Inner Hook

Figure 148.   Larva of *Tetracha carolina*. (Redrawn with permission from Hamilton 1925.)

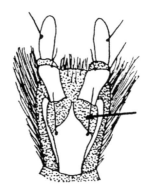

Figure 149. Labial palpiger and palpi of the third instar larva of *Tetracha carolina*. Arrow points to the membranous palpiger. (Redrawn with permission from Hamilton 1925.)

Range: A single specimen of this beetle has been reported from Bridgeport, CT (specimen is in the Peabody museum at Yale University). No further specimens have been found in the area after much searching by local cicindelaphiles, suggesting that there is no established population of this species. This specimen could be a stray individual dispersing from a population further south, or it may have been mislabeled (Tom Schultz, pers. commun.; Dunn 1986). Along with the entirely green *Tetracha virginica*, it has been reported from New York and New Jersey (Leonard 1928, Bousquet and Larochelle 1993). *Tetracha virginica* is found west to Wisconsin and Nebraska, and south to Texas and Mexico (Pearson et al. 1997). *Tetracha carolina carolina* is found further south than *T. virginica*, and west to California (Pearson et al. 1997).

Notes: Unlike other tiger beetles (*Cicindela*), *T. carolina* and *T. virginica* are active at night and hide under rocks, logs, and vegetation during the day. They can be attracted to lights, where they feed on other insects. *T. carolina* has wings but has been rarely observed flying.

Cicindela trifasciata ascendens LeConte, 1851

Common Name: S-banded tiger beetle.

Adult Identification: 11 to 13 mm. Elytra brown with slender maculations and scattered greenish pits. Ventral surface metallic blue-green. Middle band very long, thin and forming an irregular S-shape. Humeral lunule ends in a hook or lump. Apical lunule curves forward into C-shaped hook. Labrum short with a single tooth and 9 setae. Ventral side sparsely covered with decumbent hairs (Leng 1902).

Similar Species: *Cicindela repanda* is superficially similar, but has a C-shaped humeral lunule without a hook, and a simpler middle band not forming an irregular S-shape as in *C. trifasciata ascendens*.

Habitat: Similar to that of *C. marginata*; salt marsh mud flats along rivers. In the south, it is also found in rice fields (Comboni and Schultz 1989).

Life History: A single adult was collected on Nantucket Island, MA on July 25, 1988 (Comboni and Schultz 1989). Details of the life cycle in our area are unknown. There are probably no established populations north of Chesapeake Bay.

Larval Identification: Hamilton (1925: 22 to 23) describes a species called "species A" that is "Probably *Cicindela trifasciatus* [sic] *sigmoidea* LeConte." Victor Shelford in San Diego collected this specimen.

As of this writing, the larva of *Cicindela trifasciata ascendens* is undescribed (Valenti 1996a).

Range: Breeding populations are found south of Chesapeake Bay (Pearson et al. 1997). Strays to the very southern shores of New England. Commonly found along the Caribbean Sea from Mexico north the Carolinas, and in the Pacific (*Cicindela trifasciata sigmoidea* LeConte, 1851) is found from Mexico north to southern California (Pearson et al. 1997). Occasionally *C. t. ascendens* is found along the New Jersey shore (Boyd 1978). In 1988 Douglas Comboni collected one specimen on Nantucket Island, MA (Comboni and Schultz 1989).

Notes: A rarity in New England. Summer coastal storms may carry dispersing adults north. *Cicindela t. ascendens* appears to have great powers of dispersal. It has been caught at lights and on oil rigs 200 miles off the coast in the Gulf of Mexico (Graves 1982, Comboni and Schultz 1989, Knisley and Schultz 1997). Pearson et al. (1997: Map 6) shows inland strays as far from the coast as Kansas and western Tennessee.

GLOSSARY

Adephaga: One of four suborders of the order Coleoptera (Beetles). Adephaga include tiger beetles, ground beetles, diving beetles, whirligig beetles, and a few other families. Adephagous beetles have their first abdominal segment divided by the hind coxae.

Anterior: The front of, or toward the head end of, the animal.

Allopatric Species: Two or more species each of which occupies a distinct, nonoverlapping geographical range.

Alveoli: Microscopic hexagonal-shaped pits on the surface of tiger beetle cuticle (see Fig. 1). The number and depth of these pits affect the color and texture of tiger beetles.

Apex: The end of a structure farthest away from the center of the body. For example, the apical end of the elytra is the rear pointy end of the wing covers. The opposite of apex is **base.** The adjectives are "basal" and "apical." The base of the pronotum is to the rear, where it borders the base of the elytron, since the line between them approximates the middle of the body.

Caudal: Toward the tail end of the animal; posterior.

Cephalon (cephalo-)**:** The head of an animal (toward the head).

Cicindela: The official name of the dominant genus of tiger beetles worldwide. Most of the species are active by day. It contains all of the species known to be established in our area.

Cicindelaphile: A person who enjoys studying tiger beetles (literally, lover of tiger beetles).

Cicindelidae: The tiger beetle taxon when considered as a family.

Cicindelinae: The tiger beetle taxon when considered as a subfamily.

Clade: A single evolutionary line; a group of taxa sharing a closer common ancestry with one another than with members of any other clade.

Clypeus: The "bridge of the nose" area on the head of a tiger beetle between the labrum and the frons, marked off from the frons by a groove (see Fig. 3).

Concave: Curved inward to form a hollow (like the inner surface of a ball).

Convex: Curved outward to form a bump (like the outer surface of a ball).

Coxa: The first leg segment. Leg segments on beetles from the body out to the tip of the leg are: coxa, trochanter, femur, tibia, tarsi (see Figs. 5 and 6).

Cupreous: Copper-colored.

Declivity: A dropping off or declining, descending slope.

Decumbent Setae: Short hairs that lie flat against the body.

Deflexed: Turned back or down.

Distal: Toward the free end of an appendage, farthest away from the body.

Diurnal: Active during the day.

Dorsal: Toward the top of an animal.

Elytra (plural; singular: **elytron**): Wing covers on the back of beetles.

Emarginate: With a cutout place or notch in the edge or margin.

Endangered: In the legal sense, "endangered" means the species has been listed as endangered and there are laws to protect the species. However, there are species that are endangered because they have a high probability of extinction in the near future if current population trends continue, and they may not be listed or protected by any laws.

Epipleura: The lateral edge of the elytra that is visible on the ventral side of an adult beetle (see Fig. 6).

Extirpated: To become locally extinct. For example, a species could become extirpated from the state of Maine, but still be found in other states.

Femur: The third leg segment, just above the "knee." Leg segments on beetles from the body out to the tip of the leg are: coxa, trochanter, femur, tibia, tarsi (see Figs. 5 and 6).

Fovea: Pit; on tiger beetles, foveae are macroscopic colorful punctures or pits on the surface of the elytra.

Frons: The face and forehead area of an adult or larval beetle. The area of the head between the eyes (see Fig. 3).

Galea: The palpus-like 2-segmented branch of the maxilla.

Genae (plural; singular: **gena**): The lower sides or "cheeks" area of the head behind the jaws and below the eye (see Fig. 4).

Glabrous: Without hairs. A glabrous surface may be smooth or rough with ridges, punctures, or projecting points, but without hairs.

Hibernaculum: A tunnel, cavity, or protected place to pass the winter in hibernation. Tiger beetle hibernacula are usually in burrows.

Humerus or **humeral:** The "shoulder" area of the elytra or wing cover, or its anterior lateral corner.

Instar: The characteristic form of a larva between molts. For example, the first instar larva is the form of the larva after it hatches from the egg and before its first molt.

Intergrade: An individual thought to be the offspring of a mating between individuals of two subspecies. Intergrades have characteristics that are intermediate between two subspecies, and are found at the boundary of the geographical ranges of the two subspecies.

Labium: The lower or most posterior section of insect mouth parts (Fig. 3 shows labial palps).

Labrum: The upper part of the mouthparts of a beetle; often called the "upper lip" (Fig. 3).

Lateral: To the side.

Lunule: Crescent-shaped maculations or markings on the elytra (see Fig. 13).

Maculations: The light color markings on the wing covers of tiger beetles. See Fig. 13 for naming conventions.

Maxilla: In beetles, the section of the mouthpart below the mandibles or jaws, and just above the labium. The maxilla is a complex appendage, including palps used to manipulate food items (see Fig. 3).

Medial: Toward the midline of the animal.

Median Hook: The outer large hook on the dorsal side of abdominal segment 5 in tiger beetle larvae. (*Note:* this term seems illogical. The explanation is that in some genera of tiger beetles there are 3 hooks on each side of abdominal segment 5. In *Cicindela* and *Tetracha* the outer of the 3 original hooks has been lost, so the middle or median hook now appears to be lateral.)

Mesal: In the direction of the midline.

Mesic: An ecological classification of a habitat or biome; referring to an area with a moderate supply of water so that moisture-loving plants and animals may flourish.

Mesothorax: The middle section of the three-part thorax of an insect. The middle pair of legs is attached to the mesothorax.

Metathorax: The posterior section of the three-part thorax of an insect. The rear pair of legs is attached to the metathorax as are the flying wings of beetles.

Microserrations: Very tiny thorn-like projections along the posterior edge of the wing covers, giving the edge a saw-toothed appearance under high magnification.

Nocturnal: Active at night.

Nominate Subspecies: The subspecies that has the species and subspecies name the same. For example *Cicindela hirticollis hirticollis* is the nominate subspecies, while *Cicindela hirticollis rhodensis* is a different subspecies.

Obtuse: At an angle greater than a right angle; not pointed.

Ocellus: An older term for a single eye (plural: **ocelli**) on an immature insect. Also called stemma (plural: **stemmata**) (Fig. 8), which is the term in current use.

Oviposit: To lay eggs.

Palpifer: A sclerite on the stipes of the maxilla that forms the base where the palpus attaches.

Palpiger: The palpus-bearing structure on the mentum of the labium. Often appears like the first segment or plate of the labial palp in larval tiger beetles. (See couplet 1 in the "Key to Larvae" and Figs. 34 and 37.)

Palpomere: A segment of a palpus, often referring to the palpus of the maxilla.

Palpus (plural: **palpi**): A segmented, finger-like structure associated with the mouthparts. There are two pairs, the maxillary and labial palpi. The maxillary palpus has three segments in most larvae (not counting the palpifer or basal segment).

Parapatric Populations: Two or more populations whose borders are contiguous, but not overlapping.

Phylogenetic: Pertaining to evolutionary relationships within and between taxa.

Post-: Behind.

Posterior: Behind or toward the tail end (caudal).

Primary Setae: The large, usually prominent hairs occurring in fixed positions on the body. They are visible on the larva when it hatches out of the egg. Example: the hairs on the larval pronotum numbered in Fig. 46 are primary setae.

Proepisternum: The sclerite on the side of the first thoracic segment, between the pronotum and the prosternum.

Pronotum: The sclerite on the dorsum or top of the first segment of the thorax.

Prosternum: The sclerite on the ventral or belly side of the first thoracic segment.

Prothorax: The front segment of the thorax where the first pair of legs originates.

Proximal: Near, or nearest to the body (for example, proximal antennal segments are those nearest to the body). The word opposite proximal is **distal.**

Recurved: Bent backwards.

Rufous: Red color.

Scape: The first segment of the antenna nearest the head.

Sclerite: A hardened plate on the otherwise soft surface of an insect.

Scutellum: A small triangle-shaped sclerite found between the anterior edges of the elytra (Fig. 5).

Secondary Setae: Small prominent hairs. On the larval pronotum, secondary setae are usually numerous and small compared to the primary setae (see *Cicindela lepida* larval pronotum, Fig. 50). Secondary setae first appear in the second instar larvae.

Setae: Relatively large stiff hairs.

Sinuate: Having an S-shaped curve.

Striae: Longitudinal grooves or rows of punctures on beetle wing covers, characteristic of ground beetles and beetle families other than tiger beetles.

Stemma: The eyes on the larva (Fig. 8), plural: **stemmata.**

Sternum: The ventral sclerite on the extreme underside of an insect's body segment. A sternite is a subdivision of a sternal plate, or one scleritized segment.

Subapical: Almost on the apex or tip.

Subsinuate: Almost shaped like a sine curve, wavy.

Synonym: One or more scientific names used to mean the same taxon. According to the zoological rules of nomenclature, the first publication of a description of a species is the official scientific name. If someone later accidentally describes that species with a different name, the new name is a synonym. For example, *Cicindela vulgaris* Say, 1818 and *Cicindela crinifrons* Casey, 1913, are both synonyms of *Cicindela tranquebarica tranquebarica* Herbst, 1806 (Bousequet and Larochelle 1993).

Tarsus (plural: **tarsi**): One of five (in adult tiger beetles) small leg segments at the tip of the leg. Leg segments on beetles from the body out to the tip of the leg are: coxa, trochanter, femur, tibia, tarsi (Figs. 5 and 6). The tarsus has a pair of claws at its tip.

Teneral: Describes a beetle that has newly emerged from the pupa and is light in color because the pigments in its exoskeleton have not had time to complete development.

Tergum: The top or dorsal part of an insect's abdominal segment. If divided, parts are termed tergites.

Tetracha: Genus of nocturnal tiger beetles with striated elytra found to the south of our area. *Tetracha = Megacephala* (see Huber 1994).

Threatened: In the legal sense, "threatened" means the species has been listed as threatened and there are laws to protect the species. However, there are unprotected species that are threatened because their populations have been steadily declining in recent years, suggesting the species may become extinct if current population trends continue. Also, a species or local population may become suddenly threatened if some human-induced change in their habitat occurs, such as the construction of a dam or shopping mall.

Tibia: The fourth leg segment out from the body, located just below the "knee." Leg segments on beetles from the body out to the tip of the leg are: coxa, trochanter, femur, tibia, tarsi (Figs. 5 and 6).

Transverse: Broader than long, or crossing at a perpendicular angle, such as ties under a railroad track.

Tribe: A unit of insect classification between family and genus levels.

Trochanter: The second leg segment out from the body. Usually the smallest segment. Leg segments on beetles from the body out to the tip of the leg are: coxa, trochanter, femur, tibia, tarsi (Figs. 5 and 6).

Truncate: Cut off short.

Ventral (ventro-): Toward the belly or the underneath side of the animal.

Vertex: The top of the head between the eyes.

REFERENCES AND
SUGGESTED READING

Beatty, D. R. and C. B. Knisley. 1982. A description of the larval stages of *Cicindela rufiventris Dejean*. *Cicindela* 14: 1–17.

Belanger, P. 1982. Notes ecologiques sur des carabidae de ma collection. *Fabreries* 9(5): 1–16.

Bousquet, Y. and A. Larochelle. 1993. Catalogue of the Geadephaga (Coleoptera: Trachypachidae, Rhysodidae, Carabidae including Cicindelini) of America North of Mexico. *Mem. Entomol. Soc. Can. — No. 167.* Ottawa.

Boyd, H. P. 1975. The overlapping ranges of *Cicindela dorsalis dorsalis* and *C. d. media*, with notes on the Calvert Cliffs area, Maryland. *Cicindela* 7: 55–59.

Boyd, H. P. 1978. The tiger beetles (Coleoptera: Cicindelidae) of New Jersey with special reference to their ecological relationships. *Trans. Am. Entomol. Soc.* 104: 191–242.

Boyd, H. P. and Associates. 1982. *Checklist of Cicindelidae — The Tiger Beetles.* Plexus Publishing, Marlton, NJ.

Boyd, H. P. and R. W. Rust. 1982. Intraspecific and geographical variations in *Cicindela dorsalis* Say (Coleoptera: Cicindelidae). Coleopt. Bull. 36: 221–239.

Chamberland, R. 1979. Liste Annotee des principaux Carabidae (Coleoptera) dans la region de Levis. *Fabreries* 6(3): 57.

Chantal, C. 1981. La faune coleopterique (Carabidae) de L'Isle D'Anticosti. *Fabreries* 7(5): 90.

Clancy, P. 1996. Stalking tigers of the beach. *Nature Conservancy* 46 (3): 8–9.

Comboni, D. J. and T. D. Schultz. 1989. New state records for two tiger beetles (Coleoptera: Cicindelidae) in southern New England. *Entomol. News* 100(4): 150–152.

Criddle, N. 1907. Habits of some Manitoba tiger beetles (Cicindelidae). *Can. Entomol.* 39: 105–114.

Crowson, R. A. 1981. *Biology of the Coleoptera.* Academic Press, London. 802 pp.

Davis, C. A. 1903. Cicindelidae of Rhode Island. *Entomol. News* 14: 270–273.

Downie, N. M. and R. H. Arnett. 1996. Beetles of Northeastern North America. Sandhill Crane Press, Gainesville, FL.

Dunn, G. A. 1979. A New Hampshire population of *Cicindela ancocisconensis* exhibiting reduced elytral maculation. *Cicindela* 11: 61–64.

Dunn, G. A. 1981. Tiger beetles of New Hampshire. *Cicindela* 13: 1–28.

Dunn, G. A. 1986. Tiger beetles of New England (Coleoptera: Cicindelidae). *Young Entomol. Soc. Q.* 3(1): 27–41.

Dunn, G. A. and D. A. Wilson. 1979. *Cicindela marginipennis* in New Hampshire. *Cicindela* 11: 49–56.

Eckstorm, F. H. 1941. Indian place-names of the Penobscot Valley and the Maine coast. *University of Maine Studies in History and Government No. 55,* November 1941 (reprinted 1960), University Press, University of Maine, Orono. 272 pp.

Erwin, T. L. 1979. Thoughts on the evolutionary history of ground beetles: hypotheses generated from comparative faunal analyses of lowland forest sites in temperate and tropical regions. Pages 539–592 *in* Erwin, T. L. et al. (Eds.), *Carabid Beetles: Their Evolution, Natural History, and Classification.* W. Junk, the Hague.

Evans, M. E. G. 1965. The feeding methods of *Cicindela hybrida* L. (Coleoptera: Cicindelidae). *Proc. R. Entomol. Soc. London* 40: 61–66.

Freitag, R. 1965. A revision of the North American species of the *Cicindela maritima* group with a study of hybridization between *Cicindela duodecimguttata* and *oregona. Quaest. Entomol.* 1: 87–170.

Freitag, R. 1972. Comments on certain *Cicindela* from Quebec and Prince Edward Island. *Cicindela* 4: 21–23.

Freitag, R. 1974. Selection for a non-genitalic mating structure in female tiger beetles of the genus *Cicindela* (Coleoptera: Cicindelidae). *Can. Entomol.* 106: 561–568.

Gaumer, G. C. 1970. The tiger beetles of Presque Isle State Park, Pennsylvania. *Cicindela* 2: 4–7.

Graves, R. C. 1963. The Cicindelidae of Michigan (Coleoptera). *Am. Mid. Nat.* 69: 492–507.

Graves, R. C. 1965. The distribution of tiger beetles in Ontario (Coleoptera: Cicindelidae). *Proc. Entomol. Soc. Ont.* 95: 63–70.

Graves, R. C. 1982. Editor's note: another record of offshore flight in *Cicindela trifasciata. Cicindela* 14: 18.

Graves, R. C. 1988. Geographic distribution of the North American tiger beetle *Cicindela hirticollis* Say. *Cicindela* 20: 1–21.

Graves, R. C. and D. L. Pearson. 1973. Tiger Beetles of Arkansas, Louisiana and Mississippi (Coleoptera: Cicindelidae). *Trans. Am. Entomol. Soc.* 99: 157–203.

Graves, R. C., M. E. Krejci and A. C. F. Graves. 1988. Geographic variation in the North American Tiger Beetle, *Cicindela hirticollis* Say, with a description of five new subspecies (Coleoptera: Cicindelidae). *Can. Entomol.* 120: 647–678.

Graves, R. C. and D. W. Brzoska. 1991. The Tiger Beetles of Ohio (Coleoptera: Cicindelidae). *Bull. Ohio Biolog. Survey.* N.S., Volume 8, Number 4. College of Biological Sciences, Ohio State University, Columbus, OH.

Hadley, N. F. 1986. The arthropod cuticle. *Sci. Am.* 254(7): 104–112.

Hamilton, C. C. 1925. Studies on the morphology, taxonomy, and ecology of the larvae of holarctic tiger-beetles (Family Cicindelidae). *Proc. U.S. Natl. Mus.* 65: 1–87.

Hanley, W. 1977. *Natural History in America.* Massachusetts Audubon Society. Quadrangle/New York Times Book Co., New York, NY.

Huber, R. L. 1986. Citational enhancements for the Boyd checklist of North American Cicindelidae. *Cicindela* 18: 53–55.

Huber, R. L. 1994. A new species of *Tetracha* from the west coast of Venezuela, with comments on genus-level nomenclature (Coleoptera: Cicindelidae). *Cicindela* 26: 49–75.

Johnson, W. N. 1989a. A new subspecies of *Cicindela patruela* from west-central Wisconsin. *Cicindela* 21: 27–32.

Johnson, W. N. 1989b. A new subspecies of *Cicindela limbata* Say from Labrador (Coleoptera: Cicindelidae). *Nat. Can.* 116: 261–266.

Knisley, C. B. undated. Final Report to Massachusetts Natural Heritage Program. A Taxonomic Description of the Larval Stages of *Cicindela puritana*, the Puritan Tiger Beetle. Department of Biology, Randolph-Macon College, Ashland, VA 23005.

Knisley, C. B., J. I. Luebke, and D. R. Beatty. 1987. Natural history and population decline of the coastal tiger beetle, *Cicindela dorsalis dorsalis* Say (Coleoptera: Cicindelidae). *Va. J. Sci.* 38: 293–303.

Knisley, C. B., T. D. Schultz, and T. H. Hasewinkel. 1990. Seasonal activity and thermoregulatory behavior of *Cicindela patruela* (Coleoptera: Cicindelidae). *Ann. Entomol. Soc. Am.* 83(5): 911–915.

Knisley, C. B. and T. D. Schultz, 1997. The Biology of Tiger Beetles and a Guide to the Species of the South Atlantic States. *Virginia Museum of Natural History Special Publication Number 5.* Virginia Museum of Natural History, Martinsville, VA.

Kryzhanovskiy, O. L. 1976. An attempt at a revised classification of the family Carabidae (Coleoptera) [in Russian]. *Entomol. Obozr.* 55: 80–91. [English translation in *Entomol. Rev.* 55(1): 56–64.]

Laliberte, J. L. 1980. Quelques especes tres interessantes de la faune coleopterique du Quebec. *Fabreries* 7(1): 8–13.

Larochelle, A. 1971. Tiger Beetles of Magdalen Islands. *Cicindela* 3: 5–7.

Larochelle, A. 1972. Cicindelidae of Quebec. *Cicindela* 4: 49–68.

Larochelle, A. 1986a. Cicindelidae from New England in the Museum of Comparative Zoology. *Cicindela* 18: 59–64.

Larochelle, A. 1986b. A concise bibliography on the geographical distribution of the Cicindelidae of North America North of Mexico. *Cicindela* 18: 17–32.

Larson, D. J. 1986. The tiger beetle, *Cicindela limbata hyperborea* LeConte, in Goose Bay, Labrador (Coleoptera: Cicindelidae). *Coleopt. Bull.* 40 (3): 249–250.

Larson, D. J. and D. W. Langor. 1982. The carabid beetles of insular Newfoundland (Coleoptera: Carabidae: Cicindelidae) — 30 years after Lindroth. *Can. Entomol.* 114: 591–597.

Lawton, J. K. 1970. A new color variant of *Cicindela patruela*. *Cicindela* 2: 1–3.

Leonard, M. D. 1928. *A List of the Insects of New York with a List of the Spiders and Certain other Allied Groups.* Cornell University Agricultural Experiment Station, Memoir 101. Cornell University, Ithaca, NY.

Leng, C. W. 1902. Revision of the Cicindelidae of Boreal America. *Trans. Am. Entomol. Soc.* 28: 93–186 with 6 plates.

Leffler, S. R. 1979. Tiger Beetles of the Pacific Northwest (Coleoptera: Cicindelidae). Ph.D. dissertation. College of Forest Resources, University of Washington, Seattle.

Lindroth, C. H. 1954a. Carabid beetles from Nova Scotia. *Can. Entomol.* 86: 299–310.

Lindroth, C. H. 1954b. Carabid beetles from eastern and southern Labrador. *Can. Entomol.* 86: 364–376.

Lindroth, C. H. 1955. The carabid beetles of Newfoundland, including the French Islands St. Pierre and Miquelon. *Opuscula Entomologica Supplementum 12*. Entomologiska Sallskapet, Lund, Sweden. 168 pp.

Lindroth, C. H. 1961–1969. The Ground Beetles (Carabidae, excl. Cicindelidae) of Canada and Alaska. *Opuscula Entomologica Supplementum*. Entomologiska Sallskapet, Lund, Sweden.

Martin, J. E. H. 1977. *Collecting, Preparing, and Preserving Insects, Mites, and Spiders.* Part 1 in series: The Insects and Arachnids of Canada. Biosystematics Research Institute, Ottawa, ON.

Murray, R. R. 1980. Systematics of *Cicindela rufiventris* Dejean, *Cicindela sedecimpunctata* Klug and *Cicindela flohri* Bates (Coleoptera: Cicindelidae). Ph.D. dissertation. Texas A and M University, College Station, TX. 287 pp.

Nagano, C. D. 1980. Population status of the tiger beetles of the genus *Cicindela* (Coleoptera: Cicindelidae) inhabiting the marine shoreline of southern California. *Atala* 8: 33–42.

Nelson, R. E. and J. R. LaBonte. 1989. Rediscovery of *Cicindela ancocisconensis* T. W. Harris and first records for *C. scutellaris lecontei* Haldeman in Maine. *Cicindela* 21: 49–54.

Palmer, M. K. 1979. Rearing tiger beetles in the laboratory. *Cicindela* 11: 1–11.

Pearson, D. L. 1988. Biology of tiger beetles. *Ann. Rev. Entomol.* 33: 123–147.

Pearson, D. L. and F. Cassola. 1992. World-wide species richness patterns of tiger beetles (Coleoptera: Cicindelidae): Indicator taxon for biodiversity and conservation studies. *Conserv. Biol.* 6(3): 376–391.

Pearson, D. L., T. G. Barraclough, and A. P. Vogler. 1997. Distributional maps for North American species of tiger beetles (Coleoptera: Cicindelidae). *Cicindela* 29: 33–84.

Rankin, P. 1996. Cicindelidae of Nova Scotia. Unpublished undergraduate paper. Dr. David McCorquodale, Advisor. University College of Cape Breton, Sydney, NS, Canada.

Roch J. 1981. Carabidae (Coleoptera) captures en 1979 a Granby (Shefford), Quebec. *Fabreries* 7(3): 56.

Roch J. 1987. Liste annotée de quelques espèces de Coléoptèrers recoltées au camp Rolland-Germain, Frelighburg (Missisquoi), Québec De 1979 à 1985. *Fabreries* 13 (3 and 4): 69.

Roch J. 1991a. Liste annotée de quelques espèces de Coléoptèrers recoltées au camp Rolland-Germain, Frelighburg (Missisquoi), Québec. *Fabreries* 16(2): 43.

Roch J. 1991b. Liste annotée de quelques espèces de coléoptères capturées à Otterburn park, division de recensement de Rouville, Québec. *Fabreries* 16(1): 25.

Schincariol, L. A. and R. Freitag. 1991. Biological character analysis, classification, and history of the North American *Cicindela splendida* Hentz group taxa (Coleoptera: Cicindelidae). *Can. Entomol.* 123: 1327–1353.

Schultz, T. D. 1988. Destructive effects of off-road vehicles on tiger beetle habitat in Central Arizona. *Cicindela* 20: 25–29.

Shelford, V. E. 1908. Life-histories and larval habits of the tiger beetles (*Cicindelidae*). *J. Linn. Soc. London* 30: 157–184, plates 23–26.

Shelford, V. E. 1913. The life history of a bee-fly (*Spongostylum anale* Say) parasite of the larva of a tiger beetle (*Cicindela scutellaris* Say var. *lecontei* Hald.). *Ann. Entomol. Soc. Am.* 6: 213–225.

Shelford, V. E. 1917. Color and color-pattern mechanism of tiger beetles. *Ill. Biolog. Monogr.* 3(4): 399–432.

Shelford, V. E. 1963. *The Ecology of North America*. University of Illinois Press, Urbana, IL.

Sikes, Derek S. 1997. *http://viceroy.eeb.uconn.edu/CTB/Generalinfo.html*. Department of Ecology and Evolutionary Biology, University of Connecticut, Storrs, CT 06269. *dss95002@uconnvm.uconn.edu.*

Spanton, T. G. 1988. The *Cicindela sylvatica* group: Geographical variation and classification of the nearctic taxa, and reconstructed phylogeny and geographical history of the species (Coleoptera: Cicindelidae). *Quaest. Entomol.* 24: 51–161.

Stamatov, J. 1972. *Cicindela dorsalis* Say endangered on northern Atlantic coast. *Cicindela* 4: 78.

U. S. Fish and Wildlife Service. 1993a. Puritan Tiger Beetle (*Cicindela puritana* G. Horn) Recovery Plan. Hadley, Massachusetts. 45 pp.

U. S. Fish and Wildlife Service. 1993b. Northeastern Beach Tiger Beetle (*Cicindela dorsalis dorsalis* Say) Recovery Plan. Hadley, Massachusetts. 50 pp.

Valenti, M. A. 1996a. Synopsis of reported larval descriptions of tiger beetles (Coleoptera: Cicindelidae) from North America north of Mexico. *Cicindela* 28: 45–52.

Valenti, M. A. 1996b. Notes on *Cicindela rufiventris* Dejean (Coleoptera: Cicindelidae) in southern New England, U.S.A. *Cicindela* 28: 32–36.

Valenti, M. and F. E. Kurczewski. 1987. External morphology of adult *Cicindela repanda* Dejean. *Cicindela* 19: 37–49.

Vaurie, P. 1950. Notes on the habitats of some North American tiger beetles. *J. N.Y. Entomol. Soc.* 58: 143–153.

Vaurie, P. 1951. Five new subspecies of tiger beetles of the genus *Cicindela* and two corrections (Coleoptera, Cicindelidae). *Am. Mus. Novitates* 1479: 1–12.

Wallis, J. B. 1961. The Cicindelidae of Canada. University of Toronto Press. Toronto, ON.

White, R. E. 1983. *A Field Guide to the Beetles of North America*. Peterson Guide No. 29. Houghton Mifflin. Boston, MA.

Willis, H. L. 1968. Artificial key to the species of *Cicindela* of North America north of Mexico (Coleoptera: Cicindelidae). *J. Kans. Entomol. Soc.* 41: 303–317.

Willis, H. L. 1980. Description of the larva of *Cicindela patruela*. *Cicindela* 12: 49–56.

Wilson, D. A. 1971. Collecting *Cicindela rufiventris hentzi* with notes on its habitat. *Cicindela* 3: 33–40.

Wilson, D. A. 1979. *Cicindela ancocisconensis* T. W. Harris. *Cicindela* 11: 33–48.

Wilson, D. A. and A. E. Brower. 1983. The Cicindelidae of Maine. *Cicindela* 15: 1–33.

Yarbrough, W. W. and C. B. Knisley. 1994. Distribution and abundance of the coastal tiger beetle *Cicindela dorsalis media* (Coleoptera: Cicindelidae), in South Carolina. *Entomol. News* 105(4): 189–194.

INDEX

NOTES